AF809159

Analysis of Failure in Fiber Polymer Laminates

Martin Knops

Analysis of Failure in Fiber Polymer Laminates

The Theory of Alfred Puck

Corrected Second Printing

Springer

Dr.-Ing. Martin Knops
REpower Systems AG
Rotorblattentwicklung
Hollesenstraße 15
24768 Rendsburg
Germany
m.knops@repower.de

ISBN 978-3-540-75764-1 e-ISBN 978-3-540-75765-8

DOI 10.1007/978-3-540-75765-8

Library of Congress Control Number: 2008921124

© 2008 Springer-Verlag Berlin Heidelberg New York

This work is subject to copyright. All rights are reserved, whether the whole or part of the material is concerned, specifically the rights of translation, reprinting, reuse of illustrations, recitation, broadcasting, reproduction on microfilm or in any other way, and storage in data banks. Duplication of this publication or parts there of is permitted only under the provisions of the German Copyright Law of September 9, 1965, in its current version, and permissions for use must always be obtained from Springer. Violations are liable for prosecution under the German Copyright Law.

The use of general descriptive names, registered names, trademarks, etc. in this publication does not imply, even in the absence of a specific statement, that such names are exempt from the relevant protective laws and regulations and therefore free for general use.

Typesetting and production: le-tex publishing services oHG, Leipzig, Germany
Coverdesign: WMXDesign GmbH, Heidelberg

Printed on acid-free paper

9 8 7 6 5 4 3 2

springer.com

Foreword

In 1948 when I began to study mechanical engineering in the ruins of the School of Engineering in Hamburg common opinion was that designing of components was not a science but an art. Since then things have changed significantly. There is no doubt that a designer of composite components needs a lot of practice and a certain "artistic" skill but he also needs a good understanding of fundamental theory. Unfortunately still some people feel that theory is far away from practice. Personally I refer to great researchers such as the famous chemist A.L. Lavoisier (1743–1794) and the well known German scientist and teacher in aerodynamics L. Prandtl (1875–1953); it was him who said "There is nothing more practical than a good theory!" Possibly this can be a modification of the much older statement of Lavoisier: "There is nothing more practical than a good hypothesis!"

An FRP-laminate is a very sophisticated multilayered structure with skillfully tailored anisotropy. Analyzing the fracture processes in FRP-laminates needs a theory which is able to describe the different fracture mechanisms fundamentally correctly. I use as a "good hypothesis" the one created by Otto Mohr (1835–1918): "The fracture limit of a material is determined by the stresses on the fracture plane." From time to time some authors have proposed to apply this hypothesis to brittle composites. This was most forcibly argued by Hashin [Hashin 1980]. But time had not yet come for that approach. That is to say a certain computing capacity has to be available to identify the orientation of the fracture plane, which is unknown a priori.

First one had experimentally to investigate whether or not a modified Mohr-Hypothesis would be acceptable for UD-composites. Parallel to my theoretical approaches I found invaluable help for the experimental task from Ph-D candidates of the Aachen Technical University. Very often I told these young colleagues about my feelings that there is an excessive production of theories but a considerable lack of reliable experimental results. Unfortunately nowadays experimental work is not appreciated as much as it should be. Nevertheless the young colleagues from Aachen have chosen the experimental problems proposed by me as the subject of their dissertation. This way they supplied the experimental verification of my theory. I am extremely grateful for this assistance and therefore I want to list the names of all the experimenters: Dirk Huybrechts (1996), Jochen Kopp (2000), Martin Knops (2003), author of this book, Oliver Fischer (2003), Martin Mannigel (2007) and Erik Kuhnel (2008)[1].

My theory has now "arrived" in practice. Also the incorporation of my theory into the Guideline VDI 2014 (Verein Deutscher Ingenieure) as well as into guidelines of Germanischer Lloyd and Det Norske Veritas shows the same acceptance. For the readers of this book, VDI 2014 Part 3 "Development of Fiber Reinforced Plastic Components, Analysis" can be a very helpful supplement. It is written throughout in German and English. The reason that my theory has been treated in VDI 2014 so profoundly and prepared for use in practice is mainly the credit of my cooperation with Mr. Günther Lutz, who is a very experienced designer and developer of FRP-Couplings and large FRP-Drive shafts for fast ships.

In the future the Composite Technology will play an important role for the development of energy saving vehicles and technologies for environment and climate protection. The author of this book is working in this field already, developing large rotor blades for wind turbines. The fact that design engineers, using composites, have an enormous freedom in the design of "custom made materials" for special needs is a fascinating experience. Nevertheless this fascination should not stop us from thinking about the ultimate usage of our products. As can be expected one can find many composites in modern weapon systems including intercontinental ballistic missiles. It should not be our goal to develop even more weapons or make them "more effective". The problem will ever be: "How can we get rid of them before everyone will have them?"

I hope that the book of Martin Knops will satisfy many interested and critical readers. I am confident that the readers will have some delightful "ahaa… experiences". For myself I hope that for quite a while I will still be privileged to stay in contact with young composite engineers. We should never stop learning! What I already know for a long time, is: Modern composites are black and not easy to be looked through but they are not at all products of "black magic"! We can understand them, if we carefully try to think about their internal construction. However thinking alone is not enough, of equal importance are smart experiments.

Immenhausen, January 2008 *Alfred Puck*

---

[1] The year in brackets is the year of the publication of the dissertation (see also References in the book)

Preface

In 2003 I was encouraged by Alfred Puck himself to write this book about his failure model and the Puck fracture criteria. The motivation was that Puck's own book from 1996 was written in German and did not cover the major progress especially in the field of validation which was achieved since. There was no comprehensive presentation of Puck's work in English language and several people asked for an English book with such content.

I had at the time just finalized my dissertation about the gradual failure process for which I had worked closely together with Alfred Puck as a senior adviser. He considered me as one of the best experts and the right person to write such a book. That was of course a high honor to me and I agreed to start the project being supported and encouraged by the management of my employer, the Institute for Plastic Processing (IKV), namely Walter Michaeli and Ernst Schmachtenberg.

Originally it was planned to publish in 2004, but as often in such projects this was a too optimistic assumption and after I changed from university to industry in early 2005 and started a family at the same time it became hard to find enough time for the project. I am grateful to Petra Jantzen from Springer who – always polite but with the right pressure – set deadlines and pushed me to work. After all, the elongation of the project offered the opportunity to implement the latest work in the field which was conducted in the recent years. Thus, value was added to the overall project. On the other hand, there are still wishes which have not been met. For sure a broader presentation of the applications of Puck's failure model would have been desirable. But now it is time for publication and this part is left out for the moment.

There is a list of people who have made this book possible. First to name is of course Alfred Puck, who teached and advised me for years and who also revised this book. Günther Lutz – one of the co-authors of the VDI2014 guideline on the development of FRP components – has been the main reviser of the text. I thank not only for this but also for the great profit I got from the work on the VDI2014.

Additional thanks go to Walter Michaeli and Ernst Schmachtenberg who encouraged me to write this book during my time at the IKV. Besides I am very grateful to Dieter Merkle and Petra Jantzen from Springer for making this book possible. Last but not least I thank my wife Claudia for her understanding and patience during all the hours I worked for this book.

I dedicate this book to my children Paula Liane and Peter Elias.

Rendsburg, January 2008 *Martin Knops*

Contents

1 Introduction

Half a century ago the first sailplanes made of glasfiber reinforced plastic (GFRP) were lifted into the air – and some of them are still in service 50 years later. Originally they incorporated the enormous potential (strength and dimensional stability) of fiber reinforced plastics (FRP). Today they prove the durability of the material.

The designers and manufacturers of these very sophisticated and highly loaded structures entered unknown territory in the 1950ties. Qualification and manufacturing techniques had to be developed and the design process needed to be adapted to the new material. The fracture behavior of FRP was mainly unknown and a proper failure analysis did not exist in the middle of the last century.

Today it seems almost incredible that not one of the major aeronautic companies but a group of enthusiastic engineering students designed and built the first FRP-planes. One of these pioneers was Alfred Puck who studied mechanical engineering at the Technical University of Darmstadt, Germany, at that time. He developed successfully design procedures, joined the local academic gliding group and got – for the rest of his life – fascinated by fiber reinforced plastics design. Consequently, he dedicated his professional life to the research of the material behavior and the development of FRP-structures. Having worked on commercial aircraft at Bölkow Development he did for the next ten years fundamental research work on strength and deformation of high strength GRP at the "German Plastics Institut (DKI)", Darmstadt. Here, he completed his PhD-thesis about stress and strength analysis of high strength glasfiber/polymer laminates.

Further stations of his career were Ciba Geigy (Switzerland and USA), where he realized within a decade a bunch of composite applications, especially in high voltage equipment, and Kassel University. Here Alfred Puck worked as Professor for Composite Technology especially on the development of highly loaded GRP-components for automotive industry. Regarding Puck's professional career, it becomes obvious that he has gained his great experience both in industry and at university which is quite unique nowadays.

Having resigned in 1989 Alfred Puck used his spare time as a pensioner for further research activities. Based on his own PhD-thesis (compare

[Puck 1969, Puck and Schneider 1969]) and inspired by an idea of Hashin [Hashin 1980] he developed a set of revolutionary fracture criteria and a refined failure model characterized by several major advantages compared to the established criteria and models. Namely, Puck's model enables the designing engineer not only to identify the different fracture modes occurring in FRP, but also to quantify the effect a detected fracture in one layer of a laminate has on the laminate as a whole. Moreover, the gradual failure process of FRP-laminates is modeled realistically.

The superior characteristics of Puck's model have been honored lately in the context of the so called "World Wide Failure Exercise". This exercise was initialized by Mike Hinton from QinetiQ (assisted by Sam Kaddour) and Peter Soden from UMIST. They were driven by the fact that nowadays composite engineers have the choice between countless failure criteria, but very few information about the capabilities and shortcomings of these criteria is available. There is simply a great lack of experimental verification.

Hinton and Soden asked the leading experts in failure analysis to calculate with their own model stress/strain-curves and failure limits for quite a lot of test cases. The results were published in 1998 in a special issue of *Composites Science and Technology* (Soden et al. 1998). Then the organizers of the exercise presented the experimental results for all test cases and the participants were asked to comment on the difference between prediction and experiment. Simultaneously the organizers analyzed the results themselves and published them (together with the comments and explanations of the participants) in another special issue (Hinton et al. 2002). Here Puck was declared winner of the exercise.

The organizers of the exercise stress in their summarizing evaluation that Puck's model not just predicts fracture limits best, but – even more important – describes the failure phenomena taking place in the material most realistically. Using Puck's model the composite engineer can follow the gradual failure process in a laminate and deduce from the results of the analysis how to improve the laminate design. This capability distinguishes the Puck model from phenomenological and global models like the well known Tsai/Wu model (Tsai and Wu 1971; Liu and Tsai 1998).

Another result of the "World Wide Failure Exercise" is that models using fracture mechanics are today just capable to solve very few problems. Within the exercise they could just be applied to a limited set of test cases and proved not to be a helpful tool for designers at the time being (Hinton et al. 2002). However, major progress is being made in this field and the situation might change within the next decades.

The Failure Exercise concentrates on monotonously increasing loads, where the load is applied for the first time and just once and that its magnitude grows within minutes by a constant rate to its maximum value. Real structures, however, are mostly stressed by alternating loads and failure often occurs as a consequence of gradually increasing damage under oscillating load.

Nowadays, a truly reliable life time prediction is still not possible. Consequently, time consuming and expensive component tests are inevitable and iteration loops in the design process are a common problem. By conducting an adequate failure analysis with the Puck model such loops can be reduced to a minimum. This is again due to the capability of the model to distinguish the different modes of fracture and to calculate the gradual failure process taking place at the first loading of the laminate correctly.

With this knowledge the laminate's response to oscillating loads can be optimized as long as the same failure mechanisms take place as under static loads. Thus, the gradual damage process can in principle and qualitatively be evaluated by means of first loading failure analysis. Of course, such an approach does not exactly predict the load cycles to failure. Puck and Garbe have done pioneering experimental and theoretical work concerning this problem (Puck and Garbe 1993).

Both in the aeronautic and the automotive industry composites are more and more used for crash structures. Well known examples are the monocoques of Formula 1 cars and the floors of modern helicopters. Even some commercial cars like the BMW M3 already use composite crash-boxes (Fig. 1). The main advantage of FRP-structures for crash applications is their extraordinary ability to absorb energy by material destruction. However, the main disadvantage is that this process cannot be simulated realistically up to now. Especially in the automotive industry this is often a K.O.-criterion. Here, the recent progress in hardware power and crash-simulation software has reduced the development time for new cars drastically. Nowadays, far less prototypes are needed than a decade ago. In this context, new materials can just be used in automotive applications if their crash-behavior can be simulated. The pure fact that composites offer a superior crash behavior is not valid as long as it cannot be simulated.

At the moment there are several reasons for this short-coming. One is that the complex failure process taking place under crash conditions is three-dimensional and cannot be calculated using 2D-shell-elements which are – up to now – exclusively used for crash-simulation. More severe however is the fact that up to now no comprehensive model is available for the material destruction under crash-conditions. Here again the Puck model – being capable of modeling all relevant fracture modes – is a good basis for a gradual solution of the problem.

These introductory remarks are meant to illustrate the necessity and the value of a failure model which distinguishes the relevant fracture mechanisms and modes and which on the other hand is practicable for design

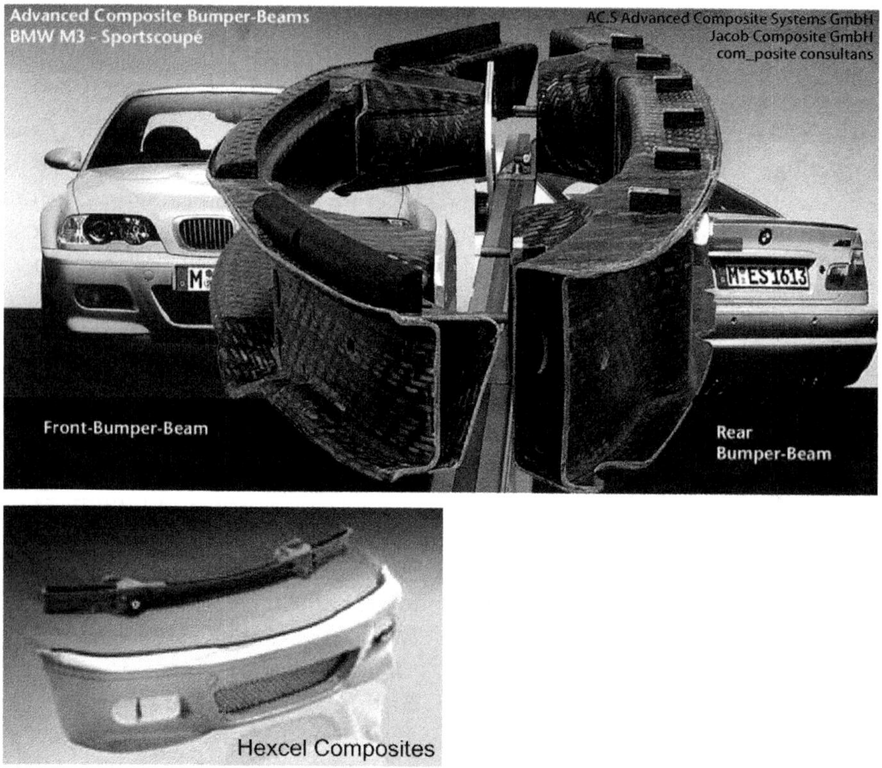

Fig. 1. Composite bumper beam of the BMW M3

engineers, a point which micro mechanical and fracture mechanical approaches do not meet today.

The first chapters of this book cover all the basics needed for a more profound discussion of failure models. It starts with the phenomena of fracture occurring in composite laminates going on to some remarks on stress analysis and finishing with failure analysis.

Briefly, some comments on the history of failure criteria are made and the different approaches to the topic are presented. These annotations prepare the ground for a comprehensive discussion of Puck's model for failure in composites. This is the main part of this book and includes the mechanical background, the mathematical formulation of the fracture criteria and the post failure model. Besides, the experimental verification and calibration mainly conducted at the Institute of Plastics Processing (IKV) in Aachen, Germany, is presented in detail. This systematical verification is unique in the history of FRP-failure-criteria and proves the maturity of the model. In an additional chapter the implementation of the theory in software is presented.

2 Failure of laminates

2.1 Laminate structure

Light-weight constructions loaded in just one constant direction are very rare. Instead the load direction might change in service and in addition the state of stress is two- or three-dimensional. Thus, several load cases are relevant for the design. Consequently, – in order to get a stable fiber network – fibers in composite parts should be placed in at least three directions.

The high potential strength of the fibers can best be exploited if no fiber ondulation occurs in the laminate. Thus, for aerospace applications and other high-end products very often unidirectional prepregs[1] are used. These are semi-finished thin sheets of pre-impregnated straight fibers with just one fiber direction, a fiber volume fraction of 60 % or more and a thickness of approximately 0.1 mm to 0.2 mm per ply[2]. Several plies with identical fiber direction are addressed as one layer[3]. A laminate is usually built up of several layers with different fiber direction.

The direction of each ply can be chosen individually, consolidation and curing of the laminate are usually conducted in an autoclave at for example 180 °C and a pressure of 6 bar. This process guaranties best product quality and highest mechanical properties. However, both the semi-finished prepregs and the process itself are rather expensive.

Alternatively, non-woven fabrics consisting of several laminae with unidirectionally oriented fibers can be impregnated with the matrix during the moulding process[4] (Fig. 2). With these non-woven fabrics about 90 % of the stiffness and strength known from the prepreg-technology are

---

1 Pre-impregnated fibers

2 thicker prepregs with up to 1 mm thickness exist, too

3 For the numerical modeling (stress-, strain- and failure analysis) such a layer is regarded as a homogeneous anisotropic continuum and addressed as a “lamina”.

4 Depending on the part size the impregnation takes place in a closed mould under a pressure of up to 6 bar (Resin Transfer Moulding, RTM) or in an open mould under vacuum assistance (Resin Infusion, RI)

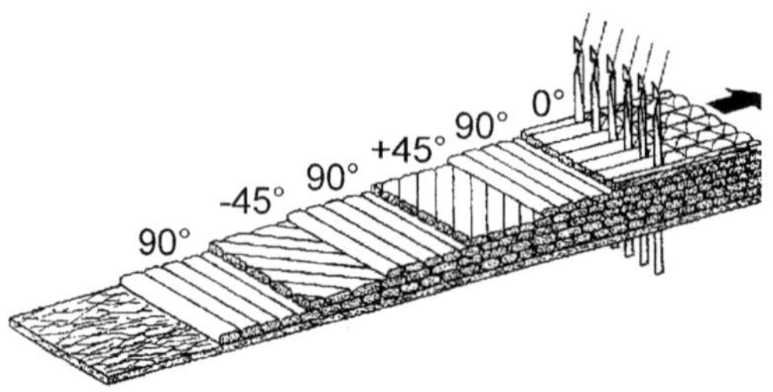

Fig. 2. Non woven fabric

reached at much lower costs. A major disadvantage of non woven fabrics is that they are rather difficult to drape. Complex geometric shapes cannot easily be reproduced. This is why – in spite of the lower strength (due to ondulated fibers) – woven fabrics are still widely used for industrial applications, mainly in prepreg form. The mechanical properties of this semi-finished product have been improved by special weaving technologies. Fabrics with a high warp/weft-ratio show only a moderate ondulation of the warp-fibers. Such fabrics are for instance used for the wings of sail-planes (Fig. 2). Here the outer shell has to guarantee the necessary stiffness against torsion and should consist of fibers orientated in ±45° to the span direction. Instead of one ply of a common fabric with a warp/weft-ratio of one, two crossing layers of fabrics mainly consisting of warp fibers are used.

A completely different technology is used for the production of rotationally symmetric parts like pressure vessels, tubes and torsion springs. For these products the filament winding technology is applied.

This brief survey illustrates that generally speaking highly loaded laminates consist of (almost) unidirectional thin layers. Fiber direction changes from layer to layer and layers with the same fiber direction are distributed homogeneously in the laminate[5]. Thus, it seems quite reasonable to model high strength laminates as a set of unidirectional laminae.

2.2 Micro-cracks

Due to different thermal expansion factors of fiber and matrix (and due to the matrix shrinkage during polymerization) considerable residual tensile

5 However, if bending prevails all fibers orientated in bending direction might be placed at the outside (e.g. top and bottom) of the laminate. The same is due for torsion (here ±45°-layers are concerned).

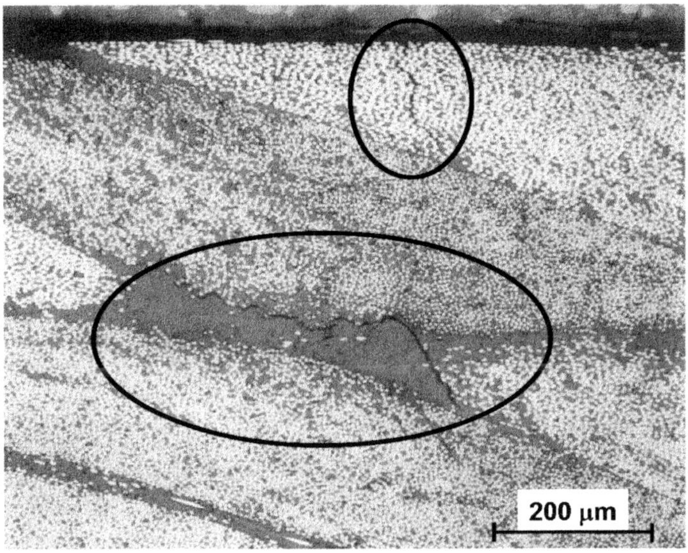

Fig. 3. Micro-cracks

stresses exist in the matrix right after curing. These stresses can lead to first small matrix cracks and local debonding of fiber and matrix. When the part is mechanically loaded over a certain limit for the first time further tiny cracks occur. As long as these cracks are non-visible and limited to a fiber/matrix-scale they are called micro-cracks. Figure 3 shows such cracks.

Both, length and number of cracks grow if the stress level in the matrix is increased. Eventually a macroscopic crack is formed which runs through the thickness of a layer. This macro-mechanical damage is called Inter Fiber Fracture (IFF) and will be discussed in the next section.

2.3 Inter Fiber Fracture (IFF)

Figure 4 illustrates a macroscopic crack of a UD-layer being embedded within a laminate. Such a crack which runs parallel to the fibers through the entire thickness of a layer is called Inter Fiber Fracture (IFF). The term IFF comprises both cohesive matrix fracture and adhesive fracture of the fiber/matrix-interface.

The crack itself does not grow slowly but is generated spontaneously and stopped only by the fibers of the neighboring layers of different fiber

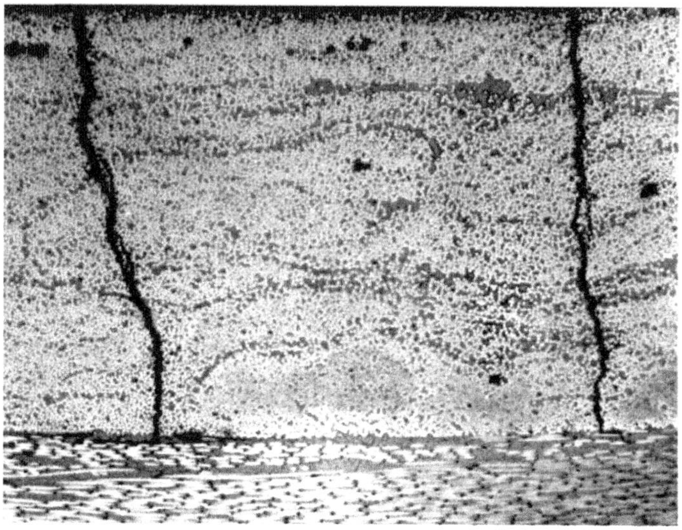

Fig. 4. Inter Fiber Fracture (IFF)

direction. Typically, at the tip of the IFF crack small delaminated zones can be observed. The cohesion between the broken and the neighboring layers is affected locally but the integrity of the laminate is still preserved. It does not fall into parts as it would naturally be the case if the laminate consisted of layers with only one fiber direction.

However, locally – transverse to the crack surfaces – the broken layer can no longer be stressed in tension. On the other hand, compressive stresses and shear stresses can still be transmitted to a certain degree due to contact and friction on the crack surfaces.

Effectively, IFF leads to a redistribution of forces within the laminate. However, it would be wrong to assume that a layer which has been damaged by IFF can – as a whole – no longer carry any load transverse to the fiber direction. In fact, at a distance from the crack the load is again introduced into the broken layer by interlaminar shear stresses (compare Fig. 5). Thus, from a macroscopic point of view the broken layer can be regarded as a continuum with reduced stiffness (Young's modulus transverse to the fibers and shear modulus).

If the load is increased after the first occurrence of IFF, further IFF-cracks will arise rapidly in the respective layer. Theoretically, a new crack will be formed right in the middle between two existing cracks. Indeed, a quite regular crack-pattern develops although IFF is in reality initiated at flaws like voids or curing cracks and not necessarily at the location of maximum stress.

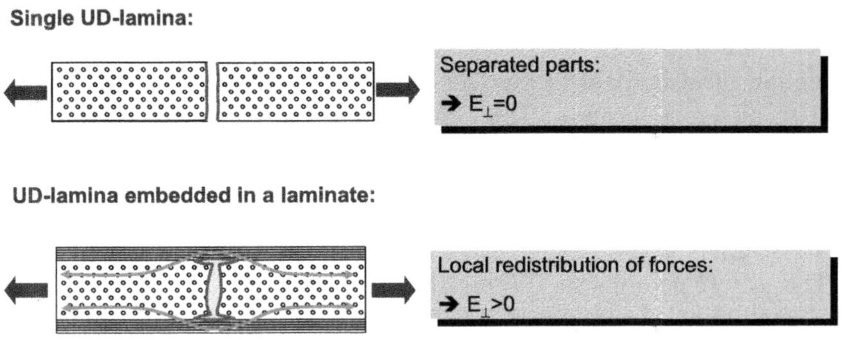

Fig. 5. Local redistribution of forces

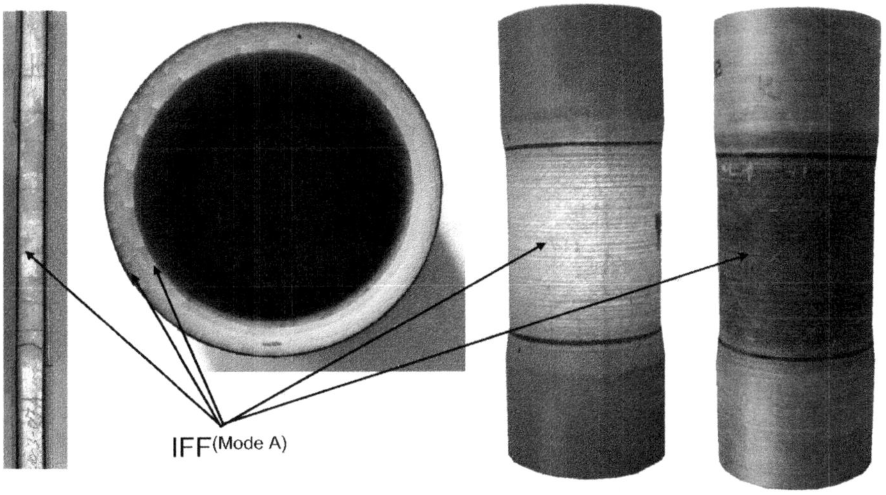

Fig. 6. Characteristic damage state (CDS)

Eventually, a so called characteristic damage state (CDS) with a maximum crack density is reached. Figure 6 shows several examples of test specimens. The mean distance between the cracks is approximately equal to the thickness of the layer. A further augmentation of the load does not lead to more cracks. Instead, the delaminated zones at the tips of the cracks will grow. This effect can be explained with energetic considerations and by means of fracture mechanics (Wittel et al. 2003).

2.3.1 Different forms of Inter Fiber Fracture (IFF)

Attention must be paid to the fact that different combinations of stresses lead to different forms of IFF. This is of major importance since some forms of IFF are harmless for certain applications whereas other forms lead inevitably to structural failure. This aspect will be discussed in more detail in the following paragraphs and in the chapter "Inter fiber fracture (IFF) criteria".

The different forms of IFF are illustrated in Fig. 7. Transverse tension and longitudinal shear both lead to straight cracks orientated parallel to the plane the stress is acting in. The plane the stress is acting in is called "action plane" and this term is of great importance in Puck's theory.

Such IFF-cracks are well known from pressure vessels. The cracks arise at the very first pressurization of the vessel. They affect the leak tightness – which must therefore be guaranteed by a thermoplastic or aluminum liner – but not the burst pressure of the vessel and in most cases also not the durability. Thus, these IFF are a good example of tolerable damage. For other applications like ultracentrifuges, however, the same damage might be classified as non-tolerable. This might be explained either by a zero-damage policy or the fact that IFF must – under cyclic loading – generally be regarded as initial damage provoking further damage and consequently reducing durability.

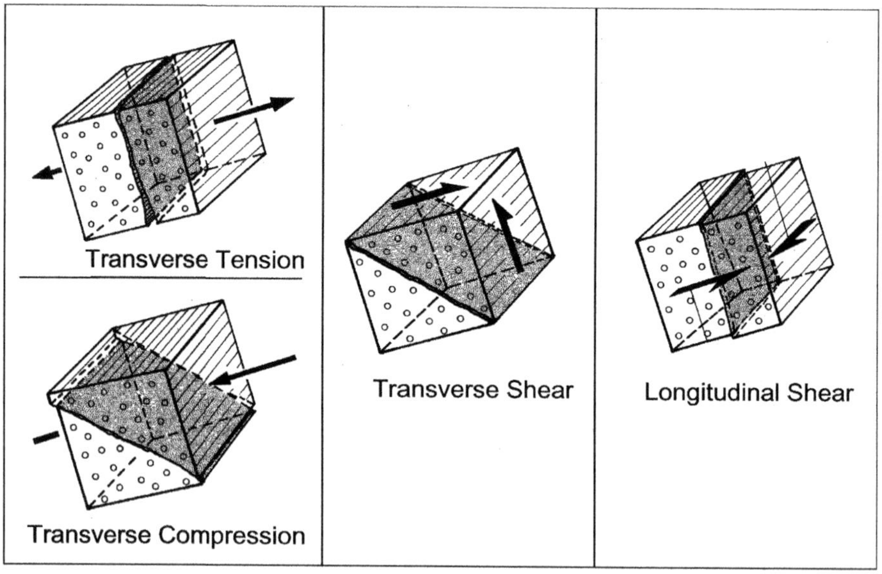

Fig. 7. Forms of IFF

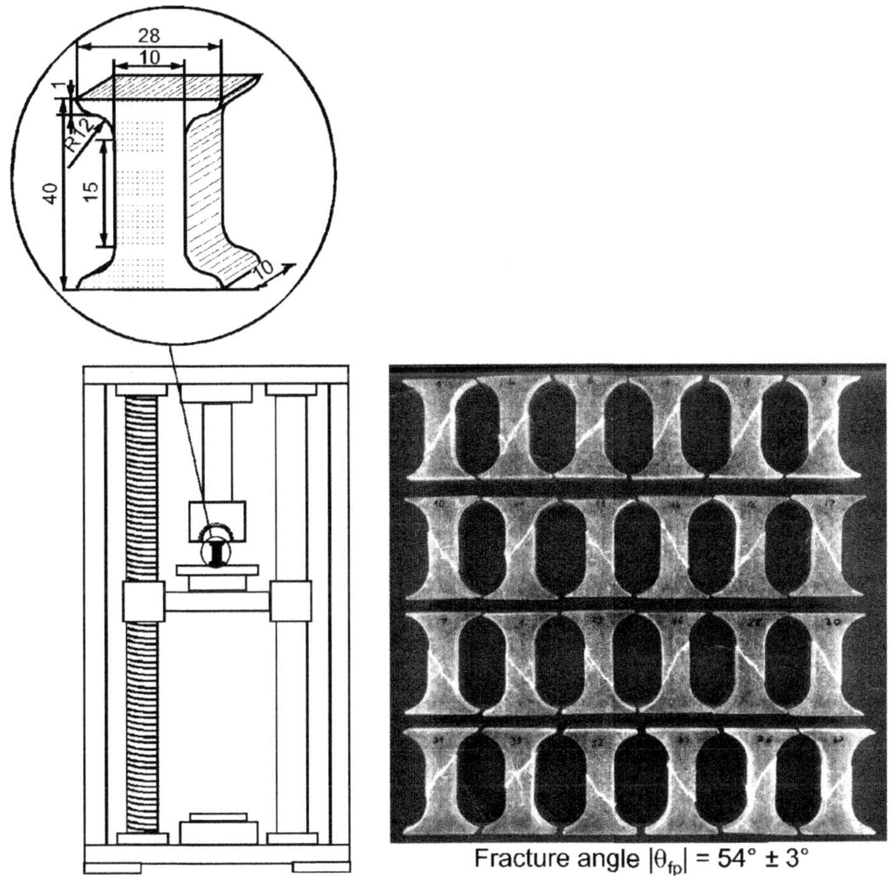

Fig. 8. Experimentally determined fracture angle under uniaxial transverse compression of UD-GFRP

In Fig. 7 the fracture surfaces caused by transverse shear and transverse compression respectively are illustrated, too. Both stressings lead to inclined fracture surfaces[6]. The inclination of the "transverse shear fracture surfaces" is exactly 45°; the inclination belonging to a transverse compressive fracture is slightly different (about ±54°, compare Fig. 8). This fracture behavior is well known from other brittle materials like concrete and grey cast iron.

Such "oblique" IFF of a layer in a laminate do not totally destroy the layer affected, but are a serious harm to the laminate as a whole. The fracture surfaces form a wedge and induce a force in thickness direction lead-

6 Inclined with respect to the plane of the applied stresses.

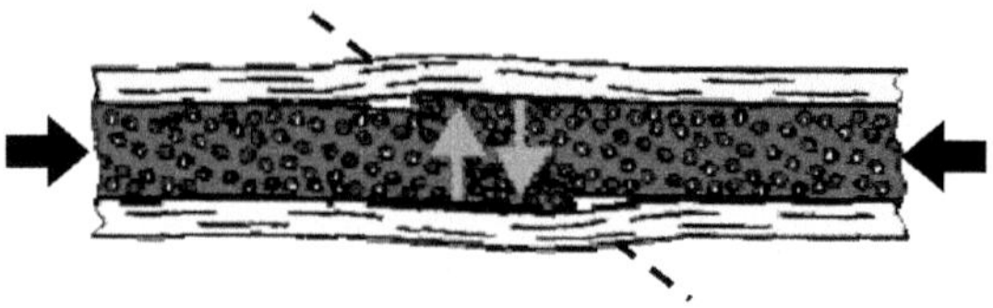

Fig. 9. Wedge effect

ing to major delamination and the dismantling of the laminate (compare Fig. 9). Thus, unlike most IFF, this form of IFF must often be regarded as final failure of the laminate. Structures of practical relevance which are potentially affected by such inclined IFF are torsional springs and vessels under external pressure like deep submerge vessels.

Such structures need to be analyzed with fracture criteria which clearly discriminate the two fracture mechanisms FF and IFF and moreover differentiate the different forms of IFF. The new fracture model developed by Puck offers unique opportunities in this field.

The relevance of a correct IFF-analysis has – among others – been proved by the prototype of a short torsional spring (500 mm long) for the rear suspension of a passenger car [Puck 1996, Puck and Schürmann 2002]. The maximum angle of twist was defined to be 30°. These boundary conditions asked for an unconventional laminate design. One of the early designs tested showed already a very good fatigue performance. This prototype consisted of a thin inner –45° layer and a thicker +45° layer

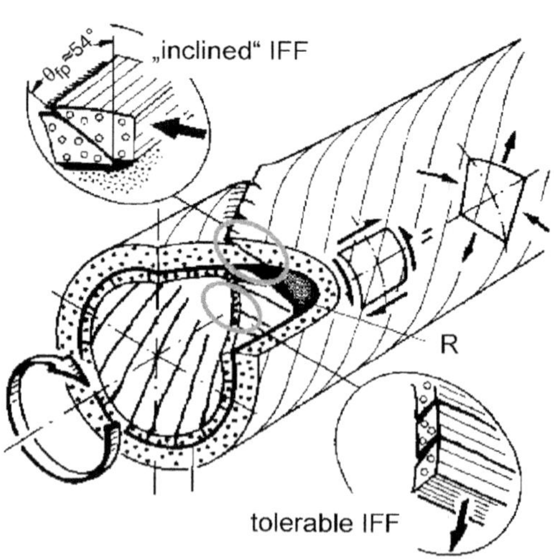

Fig. 10. Wedge effect in torsional spring

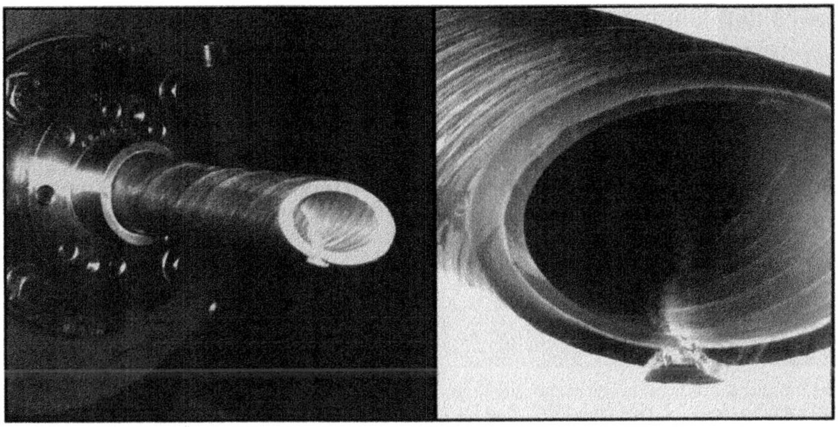

Fig. 11. Photo of wedge effect

(thickness ratio inner layer to outer layer: 3/7) both made of GFRP (glass fiber reinforced epoxy). In between these two layers a thin aramid fiber reinforced layer was added as a crack stopper. The aramid layer was used to reduce the notch effect which tolerable IFF – quickly developing in the inner lamina due to transverse tension – has on the fibers in the outer layer (Fig. 10).

Extensive testing under pulsating loading proved that multiple IFF caused by transverse tensile stress were totally harmless for the structure. Even though these IFF occurred in the inner –45° layer already after few cycles the spring survived $2{\cdot}10^6$ cycles without suffering any further visible damage. In contrast to this a single inclined IFF developing in the outer layer due to transverse compression led immediately to total failure of the torsional spring (Fig. 11). That was not a single event, but all these prototypes failed this way (compare Fig. 11).

2.3.2 Relevance of IFF

IFF caused by transverse tension and/or longitudinal shear can be tolerated for many applications. The relevance of this damage depends on the loads the laminate is subjected to and the laminate lay-up itself. Generally speaking, such IFF is harmless unless the structure under consideration is subjected to oscillating loads. In this case, the small delamination zones at the tips of the IFF can grow and lead to larger delamination zones which affect the integrity of the laminate (compare next chapter). Moreover, filaments of neighboring layers can break due to stress concentration and friction at these IFF-tips. However, the example of the torsional spring given above proves that laminates damaged by IFF can withstand millions of cycles if they are well designed.

In contrast to this, the described 'oblique' IFF – occurring under compressive stressing transverse to the fibers – cannot be tolerated. These IFF lead to spontaneous and massive delamination or even splitting of the laminate, the laminate suddenly fails due to local forces in thickness direction. This phenomenon is known as the "wedge" or "explosion" effect.

2.4 Delamination

Delamination is defined as the separation of layers from each other. This separation is caused by tensile stresses acting in thickness direction and/or shear stresses acting in planes which are parallel to the layer interfaces. Even in thin laminates under 2D-stress state such interlaminar stresses exist close to geometric discontinuities such as free edges and can be caused both by mechanical and hygrothermal loading. Even more important for the development of delamination zones are stress concentrations at inner defects such as the tips of IFF-cracks. Here high local stresses occur and cause local delamination at each IFF-crack tip. Intensive testing has been performed by Puck and his co-workers with stabilizers (torsional springs under alternating loads for trucks). Their experimental and theoretical investigations even suggest that no delamination can occur in the absence of impact as long as no IFF-cracks have developed in the laminate (Puck 1996).

Delamination zones grow under monotonously increasing or cyclic loading. Eventually two adjacent laminae can totally be separated from each other. This has often been observed on the outside of GFRP-stabilizers under high alternating loads. Delamination is a typical "crack-growth" problem and should therefore be treated by means of fracture mechanics.

In literature mainly interlaminar shear stresses are made responsible for delamination. However, it is evident that transverse tensile interlaminar stresses (so called peel stresses) provoke delamination, too, perhaps even more than interlaminar shear stresses. Apart from the obvious fact that interlaminar stresses are caused by inner defects like IFF-crack tips, these stresses are a function of the load case, curvatures of the structure and the stacking sequence. Thus, they can be reduced by a proper design of the stacking sequence [Huybrechts 1996]. A good example is the stabilizer. Here the fiber direction changes monotonously from ±45° on the inner surface to ±28° on the outside. By this means the shear stress τ_{xy} induced by the cyclic torsional moment is nearly the same for all layers. Lower winding angles at outer layers of the stabilizer also reduce the radial stresses, which are peeling stresses when they are tensile.

Ultimately the stabilizer fails gradually by delamination. The durability of the stabilizer would be enhanced if the occurrence of first IFF shifted to higher load cycles. This can for instance be achieved by the introduction of compressive residual stresses as described by Schürmann in [Schürmann 1989].

It seems necessary to mention that interlaminar shear and normal stresses do also occur in laminates which are not damaged by IFF. At free edges or cutouts interlaminar stress can reach high peaks. The reasons for interlaminar stresses which are not caused by IFF or impact can be illustrated by two simple examples illustrated in Fig. 12. Case (a) in Fig. 12 is a simple [0°/90°]-laminate under unidirectional tension in x-direction. The graphic on the upper right hand side shows the deformation as it would arise if the layers were not bonded together. Due to the different Poisson ratios $\nu_{\perp\parallel}$ and $\nu_{\parallel\perp}$ the transverse contraction of the 0°-layer is much larger than that of the 90°-layers. However, as a bonded laminate the layers deform uniformly. This leads to tensile stress σ_y in the 0°-layer and to compressive stress σ_y in the 90°-layers. At the free edges σ_y is zero and the balance of forces is restored by interlaminar shear stresses τ_{zy} which introduce the stresses σ_y into the layers.

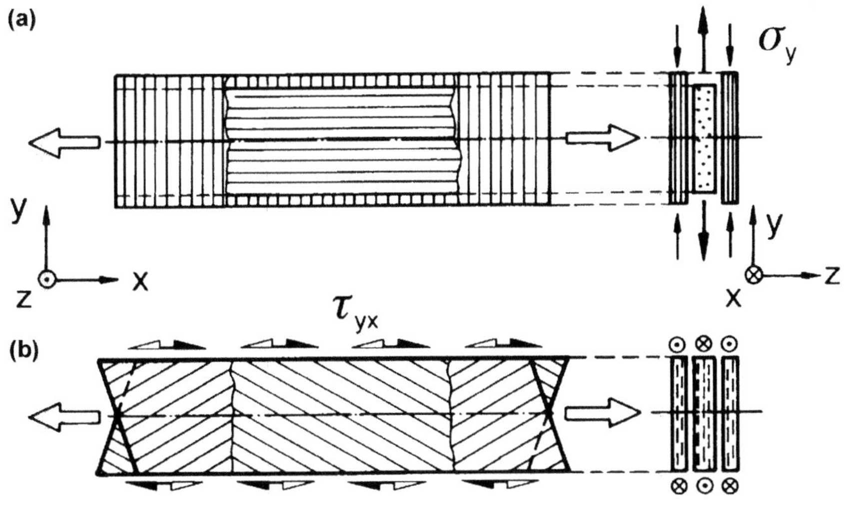

Fig. 12. Interlaminar stresses

The second example of a "balanced angle-ply" laminate with $\pm\varpi$-direction under axial tensile stress is less known. Because of their off-axis direction, the layers would undergo a shear deformation γ_{xy} of opposite sign if they were not bonded together. In the bonded laminate with zero shear strain this "natural" shear strain is suppressed by intralaminar shear stresses τ_{yx} introduced into the layers by interlaminar shear stresses τ_{zx}. These stresses reach high values (singularities) near free edges and are zero over the central region of a plate.

These brief remarks show already that a strength analysis predicting delamination is a fairly hard task, and in any case more complex than IFF-analysis. For realistic results either empirical models – taking into account the dominant influence of IFF – or fracture mechanical models need to be used.

2.5 Fiber Fracture (FF)

In a well designed laminate Fiber Fracture marks ultimate failure. It can be stated that FF is the only "desirable" fracture mechanism, since the fiber network forms the main load carrying structure. The designer should take advantage of the high strength of the fibers and eventual overloading should lead to no other fracture mechanism but Fiber Fracture. The high stiffness of the fibers results in high stresses in fiber direction. However, the stresses acting in the fibers transverse to their direction cannot be higher than those bearable by the surrounding matrix and the fiber/matrix-interface, respectively[7]. These stresses are usually by a factor of 10 smaller than the stresses acting parallel to the fibers. Thus, for isotropic glass fibers stresses transverse to the fibers and shear stresses do normally not have a significant influence on Fiber Fracture. For fibers which are anisotropic – having lower stiffness and strength in transverse direction – this might be different. Recent research work shows however, that even for carbon fibers transverse normal stresses are not relevant as far as Fiber Fracture is concerned [Michaeli et al. 2002; Fischer 2003]. The effect of shear stresses has been investigated by Mannigel [Mannigel 2007]. Mannigel found out that in CFRP shear stresses have no influence on the tensile Fiber Fracture limit. The compressive strength against FF is however

---

[7] Transverse to the fibers the unidirectional lamina can be modeled as a series arrangement of fiber and matrix. Thus nearly the same transverse stress is acting both in fibers and matrix.

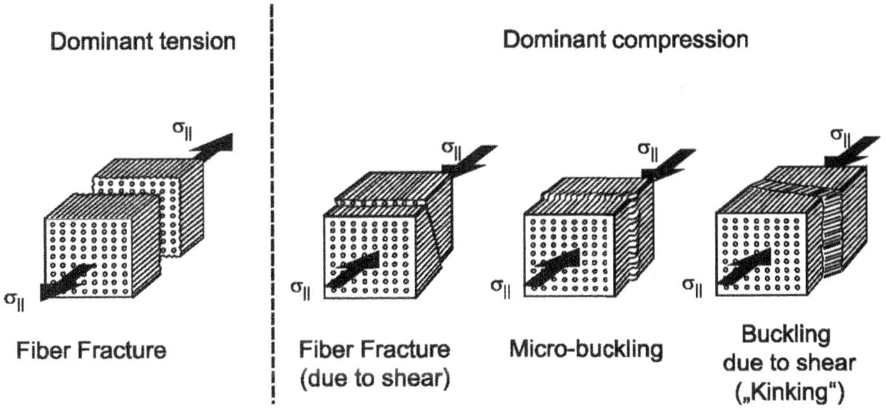

Fig. 13. Different forms of Fiber Fracture (FF)

reduced by shear stresses if these stresses are high enough to cause micro-damage and thus weaken the matrix.

It should be mentioned here, that Fiber Fracture is not defined as the rupture of single filaments but as the simultaneous fracture of several bundles of fibers effectively destroying the layer concerned in a region of macroscopic scale. Obviously, the fracture of thousands of filaments – recorded in total as Fiber Fracture – is a statistical process. Under static load some fibers will already break at some 50 % to 70 % of the Fiber Fracture limit. Under cyclic load the first filaments fail many cycles before macroscopic fracture takes place.

Figure 13 illustrates the different fracture modes occurring. Under dominant tensile stress (in fiber direction) the fibers will rupture and a failed specimen will look like a paint brush. On the other hand, under compressive stress (in fiber direction) three different fracture modes are possible.

Mostly, the layer will fail by some kind of internal buckling, called "kinking" when it takes place on a more macroscopic scale and "micro-buckling" otherwise. If carbon fibers are well aligned and the matrix supports the fibers perfectly the fibers can fail directly, without preceding buckling. In this case probably shear stresses acting in the fibers cause fracture. However, this shear fracture has seldom been observed (compare (Hart-Smith 1998a) and does probably only – if ever – occur in anisotropic fibers like carbon fibers. Figure 14 shows examples for FF under tension and compression [Fischer 2003].

If a Fiber Fracture occurs the high energy release of the failing fibers causes locally a total destruction of the fiber/matrix-composite by splitting of the fiber bundles. This prevents the layer from taking loads transverse to

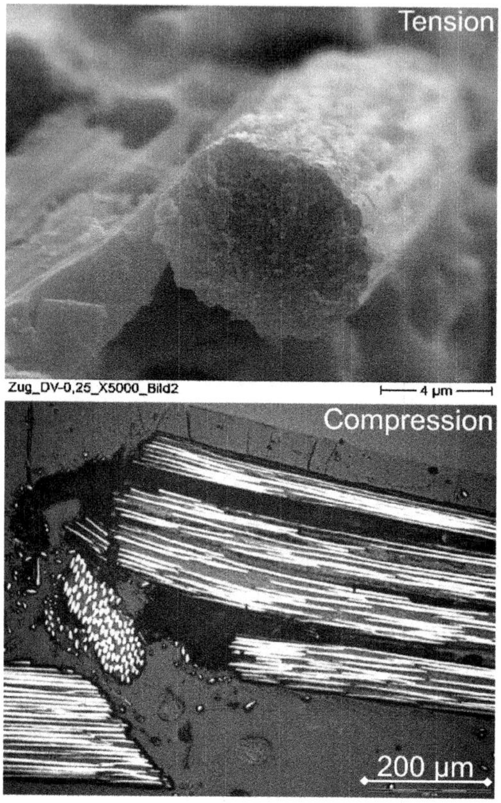

Fig. 14. Pictures of Fiber Fracture (Carbon fiber T300)

the fibers, too. It makes no sense to talk about IFF once FF has taken place. Moreover, Fiber Fracture usually leads to delamination of adjacent layers and generally speaking to damage of these layers, namely IFF or even FF, if the adjacent layer has also been highly loaded.

2.6 Laminate Failure

Obviously, a structure must be designed the way that failure does not take place during service. However, the occurrence of fracture in a layer of the laminate does not inevitably lead to laminate failure and in a well designed structure a local laminate failure will not lead to a failure of the structure as a whole. In the following paragraphs just the question of laminate failure will be discussed.

Fiber Fracture usually implies local destructions of the laminate. Even if no total failure occurs the laminate can typically no longer bear the design loads. An exception to this general statement might be simple bending beams under cycling load where some local Fiber Fracture of the outer fiber might be tolerable.

The evaluation of Inter Fiber Fracture and delamination is not as easy as that of Fiber Fracture. Delaminations may sometimes be tolerated as long as the delaminated zones remain small enough. Besides, the reason for delamination is of importance. This might be impact to the structure or Inter Fiber Fracture. As discussed above tensile IFF and the implied delamination are tolerated for many applications like pressure vessels and torsional springs. However, thoughtful design is necessary to guarantee the long term functionality of structures damaged by such tensile IFF. On the other hand, if IFF occurs as a consequence of transverse compressive stress, it implies a wedge effect and consequently extensive delamination which is usually non-tolerable and regarded as laminate failure.

2.7 Summary of chapter

In this chapter the different kinds of damage and fracture are presented. Generally, tolerable damage should be distinguished from non-tolerable damage or fracture. Whether a certain kind of damage is tolerable for a structure or not has to be defined before the design starts. Some kinds of Inter Fiber Fracture (IFF) are tolerable for many applications. For fracture analysis this means that not only Fiber Fracture (FF) and Inter Fiber Fracture (IFF) have to be clearly distinguished by the used criteria, but also the mode of IFF (including the angle of the fracture plane) must be available as a result of the fracture analysis.

3 Stress and strength analysis: Basics and definitions

3.1 Coordinate systems, stresses and stressings

3.1.1 Natural coordinate system of the UD-lamina

Figure 15 illustrates the natural coordinate system of a unidirectional lamina[8] and the global coordinate system of a laminate. The UD-lamina is an orthotropic material[9] with symmetry axes in the direction of the fibers (longitudinal), normal to the fibers in the plane of the lamina (in-plane transverse), and normal to the plane of the lamina (thickness direction). These natural axes of the "lamina coordinate system" are designated as x_1, x_2, x_3 [VDI 2006] or – in many earlier publications – simply as 1, 2, 3.

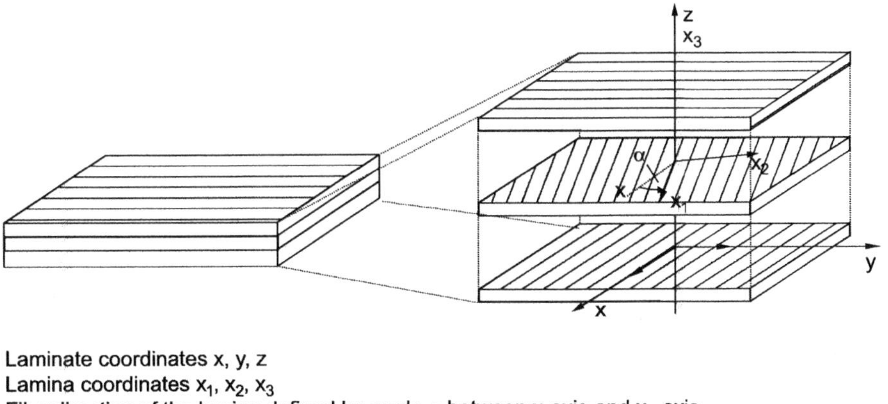

Laminate coordinates x, y, z
Lamina coordinates x_1, x_2, x_3
Fiberdirection of the lamina defined by angle α between x-axis and x_1-axis

Fig. 15. Lamina- and laminate coordinate systems

---

[8] "layer" is used for the real structure, "lamina" for the mathematical model.
[9] Defined as a material with 3 perpendicular planes of material symmetry.

3.1.2 Coordinate system of the laminate

A laminate as it is considered here is made up of two or more unidirectional laminae with different fiber directions. The laminae can be of various thicknesses and consist of different materials. Since the natural material axes differ from lamina to lamina it is more convenient to analyze laminates using a common fixed system of coordinates (x, y, z), where z is always normal to the plane of the laminate. The (x, y, z)-coordinate system is referred in the following as "laminate coordinate system". The fiber direction of a lamina is given by the angle α between the reference x-axis of the laminate and the major natural material axis (fiber direction) of the lamina, measured in a counter-clockwise direction in the (x, y)-plane [Daniel and Ishai 1994].

3.1.3 Stresses of the UD-lamina

The load acting on a laminate is usually given in the (x, y, z)-laminate coordinate system. Strains and stresses however should always be calculated in the lamina (natural) coordinate system in order to simplify the stress, strain relations by making use of the orthotropy of the UD-lamina. In the general case nine stresses exist for the lamina (Fig. 16):

- three normal stresses σ_1, σ_2 and σ_3[10],
- six shear stresses τ_{23}, τ_{31}, τ_{12}, τ_{32}, τ_{13} and τ_{21}

It is further necessary to distinguish tensile and compressive stresses. Therefore, normal stresses are often additionally marked by the index ^{t} or ^{c} indicating tensile or compressive stress respectively.

For symmetry and equilibrium reasons, corresponding pairs of shear stress are always of the same magnitude: $\tau_{21} = \tau_{12}$, $\tau_{31} = \tau_{13}$ and $\tau_{23} = \tau_{32}$. Despite of this fact, these corresponding shear stresses are consistently distinguished in this book, basically because their influence on fracture is not the same. The shear stresses τ_{12} and τ_{13} never lead to Inter Fiber Failure in the plane these stresses are acting in, since for such a fracture the strong fibers would have to be sheared off. Instead the corresponding shear stresses τ_{21} and τ_{31} are able to produce a shearing off in their fiber-parallel action plane.

The first index of the stress indicates the plane the stress is acting on. For instance, the stresses σ_3, τ_{32} and τ_{31} act on a common action plane which is

---

[10] In mechanics of composites the contracted notations (instead of σ_{11}, σ_{22}, σ_{33}) are commonly used.

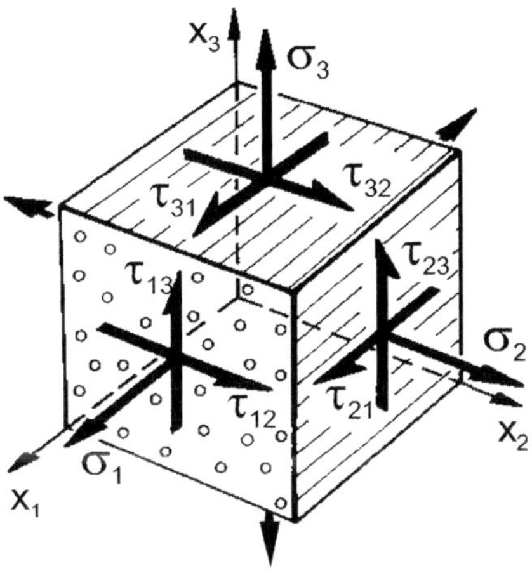

Fig. 16. Stresses of the UD-lamina

characterized by the direction of the normal vector "x_3". The second index names the direction of the stress. The shear stress τ_{32} acts in "x_2" direction, the shear stress τ_{31} acts in "x_1" direction and the stress σ_3 should correctly be named σ_{33} and acts in "x_3" direction. The second index is however left out for normal stresses, because they are always orientated in the direction of the normal vector of the plane they act on. On an action plane whose normal to the plane points towards the positive direction of the coordinate axis (also called positive section plane), a positive shear stress is acting in the direction of the positive coordinate axis (compare Fig. 16). The ultimate shear strength is – at least for a UD-lamina – the same for positive and negative shear stresses. This is a major difference to normal stresses, where the tensile and compressive strengths are not alike.

It is absolutely common practice to use the described natural coordinate system for stress and failure analysis of laminae. However, for some purposes the use of different coordinate systems or symbols is beneficial; for instance, (x_1, x_n, x_t)-coordinates are necessary for dealing with the so called action plane related fracture criteria. The definitions of these additional codes are given later.

3.1.4 Stressings of a UD-element

Quite often the use of "stressings" instead of stresses is preferable. Because of the transverse isotropy of the UD-lamina σ_2 and σ_3 stress the material in the same way. Thus, it is often helpful to talk not about stresses

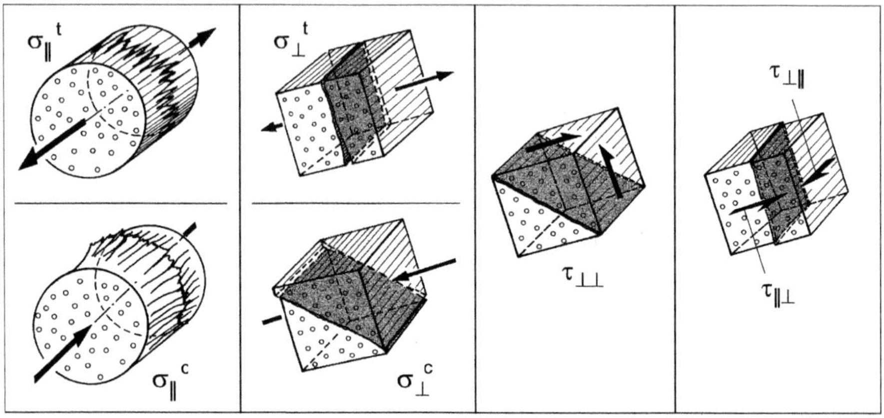

Fig. 17. Stressings

σ_2 and σ_3 but about the stressing $\sigma_\perp$ which can either be σ_2 or σ_3 (compare Fig. 17). Analogically τ_{21} and τ_{31} are addressed as longitudinal shear stressing $\tau_{\perp\parallel}$ and τ_{23} as transverse shear stressing $\tau_{\perp\perp}$. The set of stressings is completed by $\sigma_\parallel$ acting in fiber direction.

3.1.5 Stresses on the fracture plane, adapted coordinate system

In the case of IFF-analysis it is reasonable to use an adapted coordinate system. This is done because of the fact, that IFF can take place on an inclined fracture plane (compare Chapter "Inter Fiber Fracture"). Mohr has stated as early as in 1900 that in brittle material the stresses on the fracture plane are decisive for fracture under combined stressing [Mohr 1900]. These stresses on the fracture plane cannot be described using the natural coordinate system. However, this problem can easily be solved. The natural coordinate system is rotated around the x_1-axis by an angle θ[11] (Fig. 18). By doing so the stresses $\sigma_n(\theta)$, $\tau_{nt}(\theta)$, $\tau_{n1}(\theta)$ are calculated from σ_2, σ_3, τ_{23}, τ_{31}, τ_{21} using (Eq. 1) to (Eq. 3).

On the plane inclined by the angle θ (like in any other section plane) only one normal stress $\sigma_n(\theta)$ and two shear stresses $\tau_{nt}(\theta)$ and $\tau_{n1}(\theta)$ are acting (compare Fig. 18)[12]. These three stresses potentially provoke IFF on their common action plane inclined by the angle θ. This is why Puck has

[11] The angle θ is measured in a counterclockwise direction starting from the x_3-axis.

[12] The two stresses σ_t and τ_{t1} act on planes inclined by $\theta \pm 90°$.

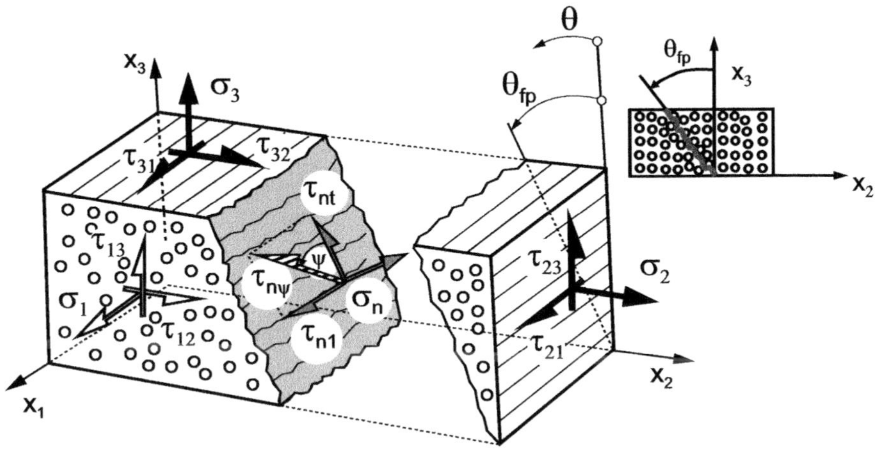

Fig. 18. Stresses of the action plane and fracture angle

formulated his IFF-criteria – which will be discussed later in detail – with the stresses $\sigma_n(\theta)$,$\tau_{nt}(\theta)$, $\tau_{n1}(\theta)$.

$$\sigma_n(\theta) = \sigma_2 \cdot \cos^2\theta + \sigma_3 \cdot \sin^2\theta + 2 \cdot \tau_{23} \cdot \sin\theta \cdot \cos\theta \qquad \text{(Eq. 1)}$$

$$\tau_{nt}(\theta) = -\sigma_2 \cdot \sin\theta \cdot \cos\theta + \sigma_3 \cdot \sin\theta \cdot \cos\theta + \tau_{23} \cdot \left(\cos^2\theta - \sin^2\theta\right) \qquad \text{(Eq. 2)}$$

$$\tau_{n1}(\theta) = \tau_{31} \cdot \sin\theta + \tau_{21} \cdot \cos\theta \qquad \text{(Eq. 3)}$$

The two shear stresses τ_{nt} and τ_{n1} can be merged to one shear stress $\tau_{n\psi}$ characterized by the angle ψ (see Fig. 18) with:

$$\tau_{n\psi} = \sqrt{\tau_{nt}^2 + \tau_{n1}^2} \qquad \text{(Eq. 4)}$$

If fracture occurs on a plane inclined by a certain angle θ this plane is called fracture plane and the angle θ is designated as θ_{fp} and called fracture angle.

In order to avoid ambiguities those stresses and stressings leading to fracture are marked with "fr". This applies both to the stresses and stressings in the natural coordinate system and the stresses on the fracture plane.

3.1.6 Coordinate system for the visualization of fracture bodies

A general state of stress given by the six stresses (σ_1, σ_2, σ_3, τ_{23}, τ_{31}, τ_{21}) cannot be visualized. This is – at first glance – just possible if stresses in

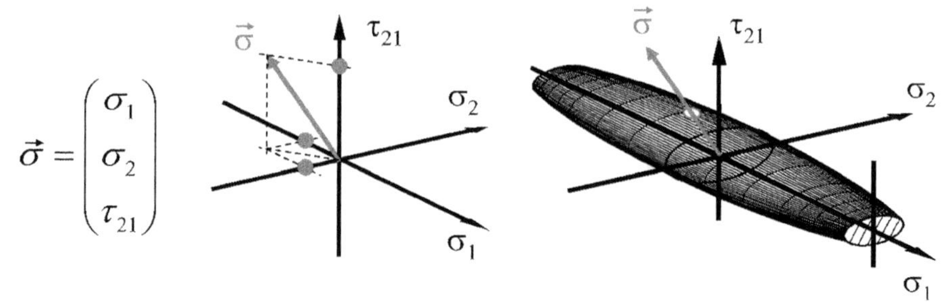

Fig. 19. Fracture body, 2D-case

thickness direction of the lamina (3-direction) can be neglected. In this case just three stresses (σ_1, σ_2, τ_{21}) remain and the state of stress can be interpreted as a stress vector with 3 components[13]. For this 2D-case a fracture condition can be interpreted as a fracture-envelope in the three-dimensional stress space defined by (σ_1, σ_2, τ_{21}) (Fig. 19). If a given stress vector penetrates the fracture-envelope, this leads to fracture. If it does not, the lamina does not fracture.

To arrive at a corresponding visualization for the general state of stress is much more difficult. This requires the combination of stresses and the definition of an adapted coordinate system. Figure 20 shows on the left hand side the stresses in the natural coordinate system (just σ_1 is missing for reasons of clearness). In a first step the stresses τ_{13} and τ_{12} can be merged to $\tau_{1\omega}$, ω being the angle between the x_2-axis and the direction of the resulting shear stress $\tau_{1\omega}$[14].

In a next step, the stresses (σ_2, σ_3, τ_{23}) are merged to (σ_{II}, σ_{III}). This procedure can be illustrated by the Mohr-circle (Fig. 20, right hand side). On planes (x_I, x_{II}) and (x_I, x_{III}) inclined by the angle φ to the (x_1, x_2)-plane and the (x_1, x_3)-plane respectively no shear stress $\tau_{II\,III}$ is acting. This elimination of shear stresses corresponds to the determination of principal stresses in isotropic materials (compare chapter Mohr's fracture hypothesis). However, in a lamina one material axis (the fiber direction designated as x_1) is fixed and the coordinate system can just be rotated around this axis. Thus, it is not useful to calculate a set of three "true" principal stresses. The stresses σ_{II}, σ_{III} are therefore correctly speaking not principal stresses.

---

13 (σ_1, σ_2, τ_{21}) are – unlike (σ_n, τ_{nt} τ_{n1}) – not acting on a common action plane! Thus, the visualization as a vector is for (σ_1, σ_2, τ_{21}) only a didactic means for illustration and better understanding.

14 Measured in a counter-clockwise direction (compare Fig. 18).

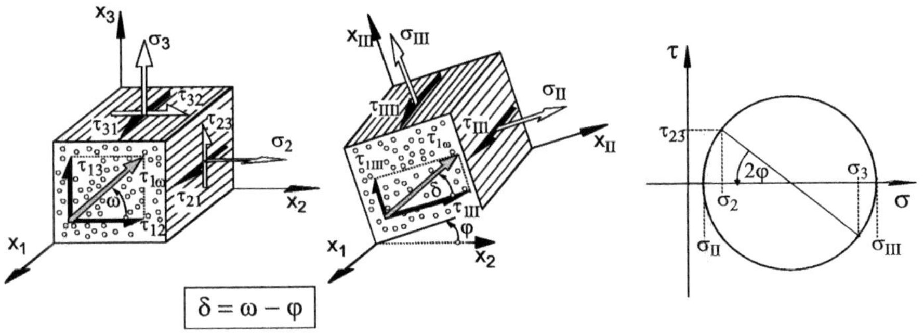

Fig. 20. Coordinate system for the visualization of 3D-fracture bodies

They are called "extreme normal stresses of the transversely isotropic (x_2, x_3) plane". In addition to these "extreme normal stresses" shear stresses τ_{II1} and τ_{III1} are present (compare Fig. 20). The corresponding shear stresses τ_{1II} and τ_{1III} can be merged just like τ_{12} and τ_{13} and the resulting stress is exactly the same ($\tau_{1\omega}$).

By the described means a set of four stresses (σ_1, σ_{II}, σ_{III}, $\tau_{1\omega}$) has been created. In addition, two angles ω and φ must be given and if the (x_{II}, x_{III})-axes are used as reference a third angle, the "angle difference" $\delta = (\omega - \varphi)$ comes into play. All this seems to be quite complicated, but it is really helpful for comparing different fracture criteria as will be shown later.

The attentive reader might be surprised to be left with four stresses, since a 4-dimensional space can still not be visualized. However, taking Mohr's hypothesis – that for brittle material only the stresses on the fracture plane provoke fracture – σ_1 is not relevant for an Inter Fiber Fracture analysis and the transverse stresses σ_{II}, σ_{III} are not relevant for Fiber Fracture. The validity of this thesis will be discussed later. At this point it is just mentioned because it allows a visualization of the complete fracture body.

3.2 Stress analysis

There are numerous publications on stress analysis in composites as for instance the textbook of Daniel [Daniel and Ishai 1994], to cite just one. Thus, in this book only some special aspects as the consideration of non-linear effects and the often neglected interaction of stresses in producing strains will be explained. In order not to deflect the reader from the main topic of Failure the chapter about the "Nonlinear stress analysis before IFF" has been moved to the Annex. Regardless of this, the chapter is more than worth reading, because the correct non-linear stress analysis is essential for a correct laminate failure analysis.

3.3 Strength analysis

3.3.1 Introduction

Deformation, strains and stresses are generally calculated using the methods of elastostatics and for many components which are designed for maximum stiffness dimensioning might be finished with the calculation of all strains. However, for typical light-weight-constructions the knowledge of stresses and strains is not sufficient. The designing engineer has to guarantee that the component will not fail in service. This implies a failure analysis and the determination of a margin of safety which might be required for technical approval of the structure. To determine the safety against failure is the task of a strength analysis or failure analysis.

Generally strength analysis correlates the load applied to the structure to the maximum bearable load. In the case of FRP-laminates a laminate lay-up which fulfills the requirements concerning the margin of safety results from strength analysis. The necessary input for a FRP-strength analysis are the stresses of the laminae resulting from stress analysis. Usually, strength analysis is done on the level of the unidirectional UD-lamina.

3.3.2 Fracture condition, fracture criterion and the term "stress exposure"

For strength analysis a mechanical-mathematical characterization of the failure process is needed. The mathematical formulation for the identification of a stress vector leading to fracture is called *fracture condition.* The general form of a fracture condition is

$$F(\sigma_j, R_j)_{j=1\,to\,n} = 1 \qquad \text{(Eq. 5)}$$

where σ_j are the components of the stress vector leading to fracture and R_j are the strength parameters for uniaxial stress or pure shear of the component j. The function F on the left hand side of (Eq. 5) is called fracture function. Generally it is a polynomial consisting of terms of the first and second power.

A *fracture criterion* distinguishes (in contrast to a fracture condition) states of stress not leading to fracture from those implying fracture. The mathematical formulation of a fracture criterion allows this clear distinction for arbitrary states of stress. The calculated stresses of the UD-lamina are – by means of a fracture criterion – correlated e. g. with the basic strengths of the UD-lamina under uniaxial stressing or pure shear stressing respectively. There are six basic strengths:

$R_{\parallel}^{t}$: fiber parallel tensile strength for uniaxial $\sigma_{\parallel}^{t}$-stressing
$R_{\parallel}^{c}$: fiber parallel compressive strength for uniaxial $\sigma_{\parallel}^{c}$-stressing
$R_{\perp}^{t}$: transverse tensile strength for uniaxial $\sigma_{\perp}^{t}$-stressing
$R_{\perp}^{c}$: transverse compressive strength for uniaxial $\sigma_{\perp}^{c}$-stressing
$R_{\perp\perp}$: transverse shear strength for pure $\tau_{\perp\perp}$-stressing
$R_{\perp\parallel}$: longitudinal shear strength for pure $\tau_{\perp\parallel}$-stressing

It is very important to realize, that all strengths are given as positive values. In contrast to the negative sign of a compressive stress the corresponding compressive strength is positive. This is a source for numerical mistakes!

The basic strengths are determined experimentally with suitable specimens. The main task of a fracture criterion is to minimize the expenses necessary for a failure analysis. Without a reliable fracture criterion for multiaxial states of stress numerous prototype tests would be needed making the design of composite parts extremely expensive and time consuming.

Usually, a fracture criterion is written as an in-equation with the stresses of the UD-lamina (σ_1, σ_2, σ_3, τ_{23}, τ_{31}, τ_{21}) as variables and the basic strengths.

$$F(\sigma_1, \sigma_2, \sigma_3, \tau_{23}, \tau_{31}, \tau_{21}, R_{\parallel}^{t}, R_{\parallel}^{c}, R_{\perp}^{t}, R_{\perp}^{c}, R_{\perp\perp}, R_{\perp\parallel}) \begin{smallmatrix} < \\ = \\ > \end{smallmatrix} 1 \qquad \text{(Eq. 6)}$$

A fracture criterion with the value "1" on the right hand side has turned to a fracture condition. If the value of the function F on the left hand side of a fracture criterion, e. g. (Eq. 6) remains smaller than "1", there is no fracture. If it is exactly "1", the fracture condition is fulfilled and fracture occurs; if the value of the function F is greater than "1", the fracture limit is exceeded. The result of (Eq. 6), i. e. a value between 0 and ∞, is frequently addressed as "failure index". The failure index does normally not supply a quantitative information about the risk of fracture. Thus, it does not tell the designing engineer to which extend he can increase the load before fracture occurs when calculating the failure index from (Eq. 6).

In order to overcome this problem one has to increase each stress component appearing in the fracture condition by a common factor, the so called stretch factor f_S such that the fracture condition becomes equal to 1, (Eq. 7). The reciprocal of f_S is f_E the so called *stress exposure*, with $f_S = 1/f_E$. From the fracture condition (Eq. 7) one can calculate f_S or f_E. However it should be noted that the fracture condition is seldomly a first order equation but rather of second order. Thus a quadratic equation must be solved in order to determine f_S or f_E respectively.

$$F(f_S \cdot \sigma_j, R_j)_{j=1\text{ to }n} = 1 \quad \text{or}$$

$$F(\frac{\sigma_j}{f_E}, R_j)_{j=1\text{ to }n} = 1 \qquad \text{(Eq. 7)}$$

It is important to note that – in contrast to (Eq. 5) – the stress components σ_j in (Eq. 7) are no more the stresses leading to fracture but they are the actual stresses, i. e. the components of the actual stress vector $\{\sigma\}$. The stress exposure f_E is therefore a direct measures for the risk of fracture caused by the actual stresses. If a given state of stress is identical to the fracture stress i. e. $\{\sigma\} = \{\sigma\}_{fr}$ then $f_S = f_E = 1$. If a state of stress $\{\sigma\}$ does not lead to fracture the stress exposure is smaller than "1" ($f_E < 1$) and the stretch factor is > 1. If the applied stress becomes to zero $f_E = 0$ and $f_S = \infty$. Therefore f_E is the preferred figure to quantify the risk of fracture.

The meaning of the term stress exposure and stretch factor can be visualized in the fracture diagram Fig. 21:

$$f_E = \frac{\text{length of the actual stress vector } \{\sigma\}}{\text{length of the vector } \{\sigma\}_{fr} \text{ of the stresses leading to fracture}}$$

$$f_S = \frac{\text{length of the vector } \{\sigma\}_{fr} \text{ of the stresses leading to fracture}}{\text{length of the actual stress vector } \{\sigma\}} \qquad \text{(Eq. 8)}$$

As briefly mentioned above in many fracture conditions not all stress terms are linear but frequently also quadratic terms exist. In such a case a value $\neq 1$ of the fracture function F of the fracture condition (Eq. 7) does **not** represent the stress exposure f_E. The result of the fracture criterion (Eq. 6)

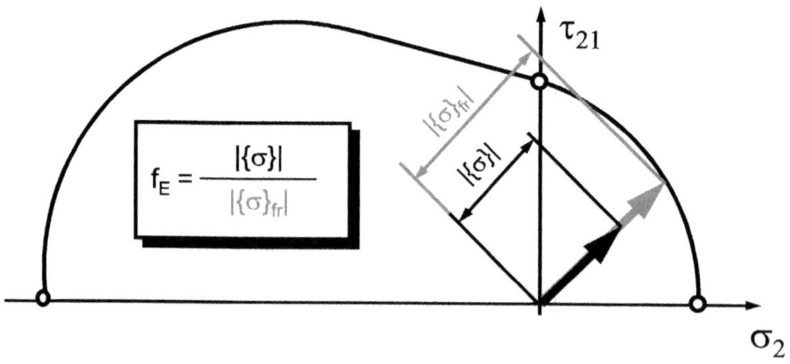

Fig. 21. Visualization of the stress exposure f_E

does only coincide with the stress exposure f_E if the fracture function F is homogenous in the first degree with respect to the stresses. Only in this special case the result of (Eq. 6) is a direct measure for the risk of fracture.

Homogenous in the first degree means that – when all stresses are multiplied by a common factor, for instance by f_S ($f_S = 1/f_E$) – this factor can be factored out. In this case, it is possible to express the effect of a multiaxial state of stress by just one value, the stress exposure f_E. The procedure can be illustrated by the example of a simple elliptical fracture condition for the interaction of the two stresses σ_2^t and τ_{21} concerning the formation of IFF:

$$\sqrt{\left(\frac{\sigma_2}{R_\perp^t}\right)^2 + \left(\frac{\tau_{21}}{R_{\perp\parallel}}\right)^2} = 1\,;\ \sigma_2 \geq 0 \qquad \text{(Eq. 9)}$$

The stresses in (Eq. 9) are the stresses leading to fracture. With the square root, the fracture function is of first order regarding the stresses and thus the fracture condition can be written with the stress exposure f_E and the actual stresses (σ_2, τ_{21}):

$$\sqrt{\left(\frac{\sigma_2}{f_E \cdot R_\perp^t}\right)^2 + \left(\frac{\tau_{21}}{f_E \cdot R_{\perp\parallel}}\right)^2} = 1\,;\ \sigma_2 \geq 0 \qquad \text{(Eq. 10)}$$

In the special case of the fracture function shown on the left side of (Eq. 9) the stress exposure f_E can be factored out in (Eq. 10) and can therefore be directly calculated from (Eq. 11):

$$\frac{1}{f_E}\sqrt{\left(\frac{\sigma_2}{R_\perp^t}\right)^2 + \left(\frac{\tau_{21}}{R_{\perp\parallel}}\right)^2} = 1\,;\ \sigma_2 \geq 0$$

$$f_E = \sqrt{\left(\frac{\sigma_2}{R_\perp^t}\right)^2 + \left(\frac{\tau_{21}}{R_{\perp\parallel}}\right)^2}\ ;\ \sigma_2 \geq 0 \qquad \text{(Eq. 11)}$$

If linear (L) and quadratic (Q) stress terms are present in the fracture function F then the stress exposure is obtained from:

$$f_E = \frac{1}{2}\left(\Sigma L + \sqrt{(\Sigma L)^2 + 4\Sigma Q}\right) \qquad \text{(Eq. 12)}$$

This equation is later on used very often when presenting Puck's action plane fracture criteria.

3.3.3 Distinguishing residual and load determined stresses

The stretch factor f_S was introduced in (Eq. 7) as a common factor of all stresses. This means that **all** stress components contributing to the stress vector $\{\sigma\}$ are magnified by the same factor $f_S = 1/f_E$. Obviously, (Eq. 7) cannot be applied in its general form, if some residual stresses are present which do not grow with increasing mechanical load. In this case the analysis is more complicated. Load-determined stresses must be distinguished from residual stresses e. g. due a temperature difference between curing and operating temperature of the structure or other reasons.

This is again illustrated in Fig. 22. In the upper part, the stress exposure f_E is illustrated for given stresses (σ_2, τ_{21}). The stress vector $\{\sigma\}$ is com-

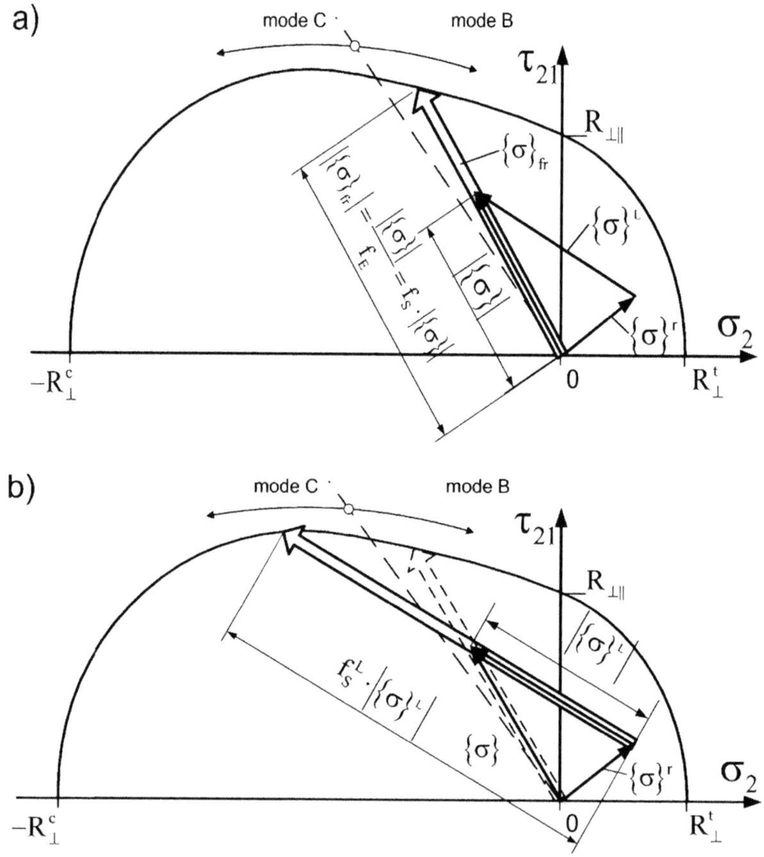

Fig. 22. Definition of the stress exposure f_E (see also Fig. 21), stretch factor f_S and stretch factor of the load-determined stresses f_S^L a) Definition of the stress exposure f_E and the stretch factor $f_S = 1/ f_E$. b) Stretch factor of the load-determined stresses f_S^L while residual stresses exist

posed of a constant residual stress vector $\{\sigma\}^r$ and a load dependent stress vector $\{\sigma\}^L$. The stress exposure is determined from the point of origin. Therefore, in this case, fracture occurs, if $\{\sigma\}$ is stretched by the factor ($1/f_E$). Thus both the load determined part and the residual stress part of the stress vector $\{\sigma\}$ is stretched by the same magnitude. In reality, however, only the load dependent stress vector $\{\sigma\}^L$ grows with increasing load whereas the residual stresses remain constant. If only $\{\sigma\}^L$ is stretched until fracture occurs, a different point of the fracture curve is hit and the magnification (stretch) factor for $\{\sigma\}^L$ is different from ($1/f_E$). The new stretch factor is called the factor f_S^L, the stretch factor of the load determined stresses.

$$F\left(\sigma_i^r + f_S^L \cdot \sigma_i^L, R_i\right) = 1 \qquad \text{(Eq. 13)}$$

In (Eq. 13) the residual stresses $\{\sigma\}^r$ are kept constant and the stretch factor for the load stresses $\{\sigma\}^L$ is searched. Obviously it is also possible to do it vice versa. This might be useful for cryogenic applications where thermal stresses are the varying load.

The stress exposure f_E and the load dependant stretch factor f_S^L are calculated for every single UD-lamina. This is done separately for IFF and FF as long as no fracture occurs. However, after IFF-initiation it makes no more sense to calculate a magnification factor f_S^L for IFF (it would be smaller than "1"). The procedure for the analysis after IFF-onset is presented in detail in the Chapter "Analysis of the gradual failure process".

3.3.4 Margin of safety and reserve factor of the composite materials structure

As a result of failure analysis the designer likes to get a factor telling him to which extend the load acting onto the structure made of composite materials can be increased before failure of the structure takes place. A better understanding of what is meant by "failure of a structure" is the approach to describe (by calculation) the incidence of a so called "limit state" of the structure. Limit states of composite structures are typically loss of tightness, loss of stiffness, loss of stability (e. g. buckling) or loss of strength by reaching the ultimate strength of the structural laminate.

A well established measure for the risk of failure or for the risk of incidence of a limit state is the reserve factor (RF). Alternatively, the so called Margin of Safety (MS) is used. A RF of "2" means that the load can be doubled before the limit state is reached. The Margin of Safety being defined as MS = RF – 1 is "1" in this case.

At first glance the definitions of both Reserve Factor and Margin of Safety seem to be trivial. However, there are many misunderstandings concerning these factors. Unlike the stress exposure f_E both RF and MF can only be defined on a laminate level of a structure and not on the lamina level. The underlying question for the definition of MF and RF is not "By which factor the current stress vector $\{\sigma\}$ of a lamina has to be increased to reach failure?" but instead "By which factor the loads on the structure and the laminate respectively have to be increased to reach failure or any limit state?". In general, that is a fundamental difference, because a doubling of the structural load might not lead to a doubling of the stresses in each lamina. This is due to non-linear material and structural behavior and stress redistribution within the laminate resulting from pre-failure damage! A second difference between stress exposure f_E[15] and Reserve Factor RF is that f_E corresponds to the overall stress vector resulting from mechanical and thermal loads along with residual stresses, whereas RF refers to the applied structural load, only. The structural load can be a mechanical load or a thermal load respectively.

In order to arrive to the RF or MS the structural load must be increased stepwise until the limit state of the structure is reached. Cases leading to residual stresses (e. g. curing of the laminate) are kept constant within such an analysis. This difference has also been highlighted in the chapter "Distinguishing residual and load determined stresses".

The calculation of RF and MS via the determination of the limit state of the composite structure can be performed in cases of statically determined structures based on analytical formulae. Typical examples are drive shafts, pressure vessels, simple plates etc. More complicated and statically undetermined structures require normally non-linear FE methods for the determination of the limit state.

3.4 Summary of chapter

In this chapter all the fundamental knowledge is presented which is necessary as a basis for the understanding of Puck's fracture criteria. These criteria will be presented in detail in the following chapter.

First the different coordinate systems used for laminate analysis are recalled and the different terms "stress", "stressing" are explained. In addition to the standard systems the coordinate system of the action plane is introduced. This is important, because with the Puck criteria the stresses of

---

[15] Or the inverse stretch factor $f_S = 1/f_E$ respectively

the action plane and not the stresses in the natural lamina coordinate system are analyzed. Moreover, a coordinate system which makes the visualization of the general 3D-state of stress possible is explained. With the help of this system fracture bodies resulting from different criteria can easily be compared.

Finally the chapter gives a general introduction into composite strength analysis. All the terms like "basic strength", "fracture criterion", "stress exposure", "stretch factor" and "margin of safety" are explained in detail. It is recommended that all readers have a look on these paragraphs in order to get familiar with the definitions which are used in the following chapters. This makes the understanding of the Puck criteria easier and avoids misunderstandings.

4 Puck's action plane fracture criteria

4.1 Fiber fracture criteria

Fiber fracture is primarily caused by a stressing $\sigma_{||}$ which acts parallel to the fibers. For (σ_1, σ_2, τ_{21})-combinations the use of a simple maximum stress formulation is recommendable. Such a formulation was already proposed by Puck in 1969 [Puck 1969, Puck and Schneider 1969]. It expresses the physical idea that fiber fracture under multiaxial stresses in a UD-lamina occurs when its stress parallel to the fibers σ_1 is equal to or exceeds the stress necessary for a fracture under uniaxial stress σ_1. From this hypothesis follows the simple FF-condition

$$\frac{\sigma_1}{R_{||}^t} = 1 \qquad \text{for } \sigma_1 > 0$$

$$\frac{\sigma_1}{\left(-R_{||}^c\right)} = 1 \qquad \text{for } \sigma_1 < 0 \qquad \text{(Eq. 14)}$$

If σ_1 is a tensile stress, the strength $R_{||}^t$ is used in (Eq. 14). If σ_1 is a compressive stress, $R_{||}^c$ is used instead. Strengths are always defined as positive values. This is the reason for using the "-" sign at $R_{||}^c$ in (Eq. 14). If σ_1 reaches the fracture stress σ_{1fr} (= $R_{||}^t$ or $-R_{||}^c$) the fracture condition is fulfilled.

(Eq. 14) can easily be transformed to a fracture criterion being formulated with the stress exposure $f_{E,FF}$:

$$f_{E,FF} = \frac{\sigma_1}{R_{||}^t} \qquad \text{for } \sigma_1 \geq 0$$

$$f_{E,FF} = \frac{\sigma_1}{(-R_{||}^c)} \qquad \text{for } \sigma_1 < 0 \qquad \text{(Eq. 15)}$$

For a preliminary analysis this simple fracture criterion is sufficient. However, for a precise analysis a more sophisticated formulation can be useful which takes into account secondary effects, too. These effects are discussed in the following:

A uniaxial σ_2-stress (or σ_3-stress, respectively) leads – due to a Poisson's effect – to an additional micro-mechanical strain in fiber direction ε_1. The effect is enlarged locally by the circumstance that the stress in the matrix is inhomogeneously distributed and effectively larger than the transverse stress σ_2 (σ_3, respectively) on the lamina-level.

This effect of stress and strain magnification can best be explained using the common model of serial springs. If a thin UD-lamina (which consists alternately of fibers and matrix) is stressed transverse to the fiber direction, the load is – on an intersection without fibers – carried by the matrix alone. On a neighboring intersection, the load is carried by the fiber alone [Puck and Schneider 1969]. From this simple model follows that the stress σ_2 is the same in fiber and matrix and that the micro-mechanical strain is different due to the different Young's moduli. However, close to the interface between fiber and matrix there is additional biaxial strain in the matrix. This effect is taken into account by the use of a magnification factor $m_{\sigma,f}$. Puck proposes $m_{\sigma,f} = 1.3$ for GFRP and $m_{\sigma,f} = 1.1$ for CFRP[16] [Puck 1996, Fischer 2003].

Puck based his more sophisticated FF-condition on the following fracture hypothesis [Puck 1996, Puck and Schürmann 1998, Puck and Schürmann 2002]:

Fiber fracture of a UD-lamina under combined stresses will occur when in the fibers the same stress σ_{1f} is reached which is acting *in the fibers* at an FF of the lamina caused by uniaxial tensile stress σ_1^t or a uniaxial compressive stress σ_1^c respectively.

The starting point is the strain ε_{1f} of the fibers caused by the combined stresses $\sigma_1, \sigma_2, \sigma_3$:

$$\varepsilon_{1f} = \frac{\sigma_{1f}}{E_{\parallel f}} - \frac{\nu_{\parallel\perp f}}{E_{\perp f}} \cdot m_{\sigma f} \cdot \left(\sigma_2 + \sigma_3\right) \qquad \text{(Eq. 16)}$$

Using $\nu_{\parallel\perp f}/E_{\perp f} = \nu_{\perp\parallel f}/E_{\parallel f}$, and $\varepsilon_{1f} = \varepsilon_1$, the stress σ_{1f} in the fibers acting in their longitudinal direction can be calculated from

$$\sigma_{1f} = E_{\parallel f} \cdot \varepsilon_1 + \nu_{\perp\parallel f} \cdot m_{\sigma f} \cdot (\sigma_2 + \sigma_3) \qquad \text{(Eq. 17)}$$

---

[16] The value for CFRP is smaller because the transverse Young's modulus of carbon fibers is lower than that of glass fibers.

From (Eq. 17) a new fracture condition can be derived. First ε_1 and the fiber stress σ_{1f} are replaced by the elastic law of the UD-lamina and the fracture resistance $R_{\parallel f}$ of the fibre, respectively:

$$\varepsilon_1 = \frac{\sigma_1}{E_\parallel} - \frac{\nu_{\perp\parallel}}{E_\parallel}(\sigma_2 + \sigma_3) \tag{Eq. 18}$$

$$R_{\parallel f} = E_{\parallel f} \cdot e_\parallel \; ; R_\parallel = E_\parallel \cdot e_\parallel$$ with $e_\parallel$ = fracture strain from uniaxial σ_1 of both fiber and UD-lamina (Eq. 19)

Based on this the FF-condition for the UD-lamina is:

$$\frac{1}{\pm R_\parallel^{t,c}}\left[\sigma_1 - \left(\nu_{\perp\parallel} - \nu_{\perp\parallel f} m_{\sigma f}\frac{E_\parallel}{E_{\parallel f}}\right)(\sigma_2 + \sigma_3)\right] = 1 \quad \text{with}$$

(Eq. 20)

$$\begin{matrix} R_\parallel^t & for\ [...] > 0 \\ -R_\parallel^c & for\ [...] < 0 \end{matrix}$$

This fracture condition (Eq. 20) is homogeneous of grade 1 and can thus easily be transformed to a formulation for the stress exposure $f_{E,FF}$:

$$f_{E,FF} = \frac{1}{\pm R_\parallel^{t,c}}\left[\sigma_1 - \left(\nu_{\perp\parallel} - \nu_{\perp\parallel f} m_{\sigma f}\frac{E_\parallel}{E_{\parallel f}}\right)(\sigma_2 + \sigma_3)\right] \quad \text{with}$$

(Eq. 21)

$$\begin{matrix} R_\parallel^t & for\ [...] \geq 0 \\ -R_\parallel^c & for\ [...] < 0 \end{matrix}$$

Here, $f_{E,FF}$ is the FF-stress exposure of the UD-lamina, $\nu_{\perp\parallel}$ is the major Poisson's ratios of the UD-lamina, and $\nu_{\perp\parallel f}$ the major Poisson's ratio of the fiber. In general the definition of the Poisson's ratios is a well known source of misunderstandings, because there are different definitions used. In the US for instance, it is common standard to define the indices of ν the other way round (the first index for the direction of the stress which leads to the Poisson effect).

In the above Equations $E_\parallel$ is the longitudinal modulus of the lamina parallel to the fibers and $E_{\parallel f}$ the longitudinal modulus of the fibers. $m_{\sigma f}$ is a magnification factor for the transverse stress in the fibers (GFRP $m_{\sigma f} \approx 1{,}3$ and for CFRP $m_{\sigma f} \approx 1{,}1$). Puck has found that in case of plane (σ_1, σ_2, τ_{21})-stress, the results of (Eq. 15) and (Eq. 21) differ only by a few percent. However, the influence of transverse stresses on FF can become important in the region of combined $\sigma_2<0$ and $\sigma_3<0$ of similar magnitude, where $|\sigma_2|$

and $|\sigma_3|$ can exceed the transverse compressive strength $R_\perp^c$ by a factor of up to 3 or 4 [Kopp 2000]).

The maximum stress criteria (Eq. 15) and (Eq. 21) presented above were put into question by Hart-Smith [Hart-Smith 1998a, Hart-Smith 1998b] who proclaimed that the fracture stress σ_{1fr} is considerably reduced in the presents of high transverse strains with a different algebraic sign compared to the strain in fiber direction ε_1. Hart-Smith based this on some experimental work [Hart-Smith 1984]. However, this hypothesis has – at least for monotonously increasing loads – been disproved by sophisticated experimental work [Fischer 2003; Mannigel 2007] which approves the appropriateness of maximum stress criteria.

A further aspect to be considered is the influence of shear stresses on the fiber fracture. Whereas tensile stressing in fiber direction $\sigma_\parallel^t$ leads unquestionably to fiber rupture, the fracture mode is not as obvious in the case of compressive stressing $\sigma_\parallel^c$. In principle, several forms of fiber fracture are possible. The dominant failure mode is kinking of the fibers due to shear stresses in the fiber [Schürmann 2004; Pinho et al 2006]. Consequently, longitudinal shear stressing of the UD-lamina $\tau_{\perp\parallel}$ reduces the compressive fracture strength $\sigma_1{}^c{}_{fr}$. In [Puck 1996] and [Fischer 2003] pure empirical formulations are given to account for this effect. Mannigel [Mannigel 2007] quantified the effect for the first time by intensive experimental investigations. He found that for relatively small shear stresses there is no influence of τ_{21} on the fiber fracture. If the shear stress surpasses a threshold and leads to microdamage, the compressive fracture strength is significantly reduced. Mannigel found a linear (degressive) influence of the shear stress on the compressive fracture strength above the threshold. A calibration can be made with the experimental results published in [Mannigel 2007].

A further influence on the fiber fracture which is not taken into account by the maximum stress criteria (Eq. 15) and (Eq. 21) is the effect IFF has. Especially under cyclic loading IFF reduces the fiber fracture stress σ_{1fr} considerably. However, the UD-lamina can in this case no longer been used as an isolated model. Instead, IFF in adjacent lamina influences FF [Knickrehm 2000]. In this case a more sophisticated analysis of the gradual failure process is necessary.

4.2 Inter fiber fracture (IFF) criteria

4.2.1 Motivation

The major motivation for the development of Puck's Inter Fiber Fracture criteria is experimental evidence. Fracture criteria are used to predict the

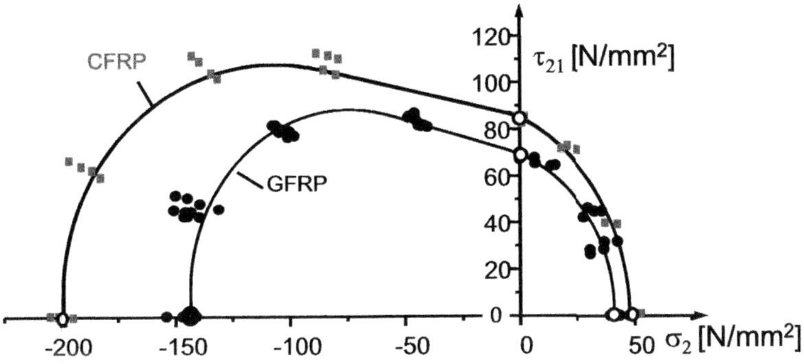

Fig. 23. Fracture limits for (σ_2, τ_{21})-stress combinations

strength of a material for arbitrary states of multiaxial stress with the preliminary knowledge of only some easy to measure strength values. One can best compare different criteria by taking experimental results that do exhibit some kind of "uncommon" or special material behavior. For carbon (squares) and glass fiber reinforced plastics (circles) Fig. 23 shows such test results obtained in a large German research project [Cuntze et al. 1997].

Some remarkable observations can be made here:

- The stresses σ_2^t and τ_{21} interact. That means that – when σ_2^t and τ_{21} act at once – fracture occurs before $\sigma_2^t = R_\perp^t$ or $\tau_{21} = R_{\perp\parallel}$ respectively are reached.
- Shear fracture is impeded by moderate transverse compressive stress σ_2^c. In other words: Higher shear stress τ_{21} can be sustained without fracture, when τ_{21} and moderate σ_2^c act simultaneously.
- A third aspect becomes obvious when looking at Fig. 24: When σ_2^c and τ_{21} act simultaneously and the ratio $|\sigma_2^c/\tau_{21}|$ exceeds a certain value, fracture occurs under an angle $\theta_{fp} \neq 0°$. The fracture angle increases with growing stress ratio $|\sigma_2^c/\tau_{21}|$ and reaches approximately ±54° for pure transverse compressive stress (see Fig. 8).

The real failure envelope is not symmetric to any vertical line. This is clearly illustrated by the fracture curve in Fig. 24 and cannot be modeled by conventional global failure criteria. Besides, these criteria cannot calculate the fracture angle nor do they explain the effects observed and explained above. However, the understanding of these effects is crucial for the development of reasonable failure criteria. Of special interest is in this context the analysis of the fracture angle under pure transverse compression σ_2^c. Fracture occurs on a plane inclined by approximately ±54° to the action plane of σ_2^c. A simple sketch of the Mohr circle shows what stresses act on this fracture plane (Fig. 25). This shall be derived in detail:

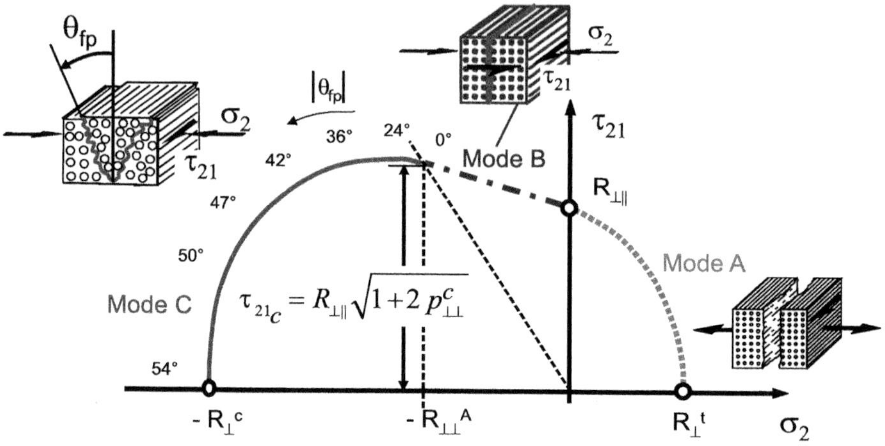

Fig. 24. Fracture curve (σ_2, τ_{21})

On the action plane of σ_2^c ($\theta = 0°$) pure transverse compression ($\sigma_n^c = \sigma_2^c$, $\tau_{nt} = 0$) is acting. On this plane no fracture can occur. On a perpendicular plane ($\theta = 90°$) there is no stress at all. On all other planes with $0° < \theta < 90°$ combinations of a transverse compressive stress $\sigma_n^c(\theta)$ and a transverse shear stress $\tau_{nt}(\theta)$ are acting. Under $\theta = 45°$ the maximum shear stress $\tau_{nt\ max} = \tau_{nt}(45°)$ is reached. It has the same value as $\sigma_n^c(45°)$ acting on the same plane and the magnitude is just half that of σ_2. Under 54° – on the fracture plane – the shear stress $\tau_{nt}(54°)$ is just slightly smaller than $\tau_{nt\ max}$. However, the transverse compressive stress $\sigma_{nt}(54°)$ which impedes fracture is considerably smaller than the corresponding value un-

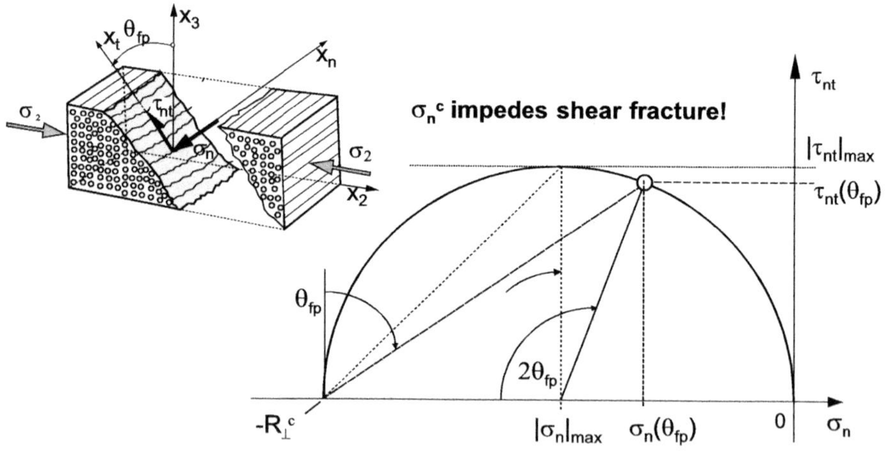

Fig. 25. Fracture under uniaxial compressive stress σ_2^c

der 45°. This fact explains why the fracture angle is greater than ±45°. This is in accordance with the observation that higher shear stresses can be sustained in the presence of moderate transverse compression.

The analysis of the fracture behavior which will be presented here, works with the stresses on the action plane and not with the lamina stresses σ_2, σ_3, τ_{23}, τ_{31}, τ_{21}. This is a fundamental fact and leads to the approach to formulate fracture criteria using the stresses of the action plane. These fracture criteria will be presented and explained in the next sections. Before this, the fracture hypotheses those criteria are based on, are discussed. Besides, the basic strengths and some inclination parameters needed for calibration are presented.

4.2.2 Different IFF-fracture modes

The different kinds of IFF described above for the (σ_2, τ_{21})-fracture curve lead to the differentiation of three IFF-fracture modes, namely Mode A, Mode B and Mode C defined as follows:

- *Mode A*: Transverse tensile stressing $\sigma_\perp^t$ or longitudinal shear stressing $\tau_{\perp\parallel}$ cause – acting either alone or in combination – fracture. In the case of a 2D-state of stress with σ_1, σ_2, τ_{21} the cracks run in thickness direction and thus in the common action plane of the lamina stresses σ_2 and τ_{21}. The fracture surfaces are separated from each other due to the tensile stressing. This leads from a macroscopic point of view to a degradation of both the Young's modulus $E_\perp$ and the shear modulus $G_{\perp\parallel}$.
- *Mode B*: Fracture is caused by longitudinal shear stressing $\tau_{\perp\parallel}$. This fracture occurs on the action plane of the external shear stress τ_{21}. In contrast to Mode A, the transverse normal stressing $\sigma_\perp^c$ which acts on the fracture plane simultaneously with $\tau_{\perp\parallel}$ is a compressive stressing! Thus the crack does not open and the fracture surfaces are pressed on each other[17]. Consequently the degradation of stiffness due to $IFF^{Mode\ B}$ is much less significant than that due to $IFF^{Mode\ A}$. An $IFF^{Mode\ B}$ occurs as long as the ratio of the compressive stress at fracture and the transverse compressive strength $|\sigma_{\perp fr}^c/R_\perp^c|$ is smaller than roughly 0.4.
- *Mode C*: If the ratio of the compressive normal stressing at fracture and the transverse compressive strength $|\sigma_{\perp fr}^c/R_\perp^c|$ exceeds roughly 0.4, the action plane of the external shear stress τ_{21} is no longer the fracture plane. Instead fracture occurs on a plane inclined by an angle $|\theta_{fp}| \neq 0°$ to the action plane of σ_2 and τ_{21}. The fracture angle $|\theta_{fp}|$ increases in the case of a plane state of stress (σ_1, σ_2, τ_{21}) from 0° at the threshold

[17] Thermal tensile stresses in the matrix might lead to a separation of the fracture surfaces nonetheless.

between Mode B and Mode C to roughly |54°| for pure transverse compression ($\tau_{21} = 0$). $IFF^{Mode\ C}$ implies the risk of delamination between the broken layer and adjacent layers.

In a more general 3D state of stress a forth fracture mode can occur:

- *Mode A**: On the fracture plan acts a combination of $\sigma_\perp^t$, $\tau_{\perp\perp}$ and $\tau_{\perp\parallel}$.

4.2.3 Fracture hypotheses

4.2.3.1 Mohr's fracture hypothesis

Fiber fracture (FF) is – in the sense as it used here – defined as the simultaneous breaking of thousands of filaments. FF of a UD-lamina is regarded as final failure of the affected lamina. Inter fiber fracture (IFF) on the other hand is defined as a macroscopic crack which runs parallel to the fibers and separates an isolated UD-layer into two pieces. Usually, such a macroscopic separation is preceded by micro-mechanical damage of the matrix or the fiber/matrix-interface, respectively.

Fiber reinforced composites (FRP) used in lightweight construction usually show brittle fracture behavior concerning both FF and IFF. This means that fracture occurs suddenly on a certain fracture plane without any major plastic deformation. This characteristic material behavior is especially distinctive for $\sigma_\perp^t$- and $\tau_{\perp\perp}$-stressing. In both cases brittle fracture occurs on the plane with the highest tensile stress (compare Fig. 26). Paul [Paul 1961] called materials with such a behavior "intrinsically brittle". The compressive strength of such materials (here: $R_\perp^c$) is much higher than (usually twice the value of) the tensile strength (here: $R_\perp^t$). Uniaxial compression leads to shearing on a plane which is inclined to the direction of the compressive stressing (compare Fig. 26). This fracture is caused by the transverse shear stressing (here: $\tau_{\perp\perp}$) acting on the inclined fracture plane.

The described brittle fracture in FRP-structures has been known for decades now. Nevertheless, most commercial Finite-Element-programs still work with global strength criteria like that of Tsai/Wu [Tsai 1992], which are in principle all based on the von Mises yielding criterion. Those global strength criteria can only be regarded as pure interpolating formulae. They are not based on a failure hypothesis and thus do not consider the true material behavior. All stresses are put into one formula not regarding whether the single stress causes FF or IFF.

Obviously, yielding criteria are not adequate for FRP-failure analysis. Because of the brittle material behavior of FRP, failure analysis should instead be based on the ideas of Mohr [Mohr 1900] and Coulomb

[Coulomb 1776]. **Mohr's fracture hypothesis** for brittle materials is the following:

> *The **fracture limit** of a material is **determined by** the **stresses on the fracture plane**.*

The cited original work refers to brittle isotropic material (for instance cast iron).

4.2.3.2 Puck's fracture hypotheses

Most failure conditions in use are pure mathematical interpolation functions with no physical basis. Puck, however, based his fracture criteria on a reasonable understanding of the brittle fracture, formulated in his fracture hypotheses. He works with the stresses σ_n, τ_{nt} and τ_{n1} which are acting on a common fiber parallel action plane of a UD-lamina. The two shear stresses τ_{nt} and τ_{n1} can be combined to one resulting shear stress $\tau_{n\psi}$ (see (Eq. 4) and Fig. 18). Puck's fracture hypotheses are the following:

1. Inter Fiber Fracture on a plane parallel to the fibers is caused by the stresses σ_n and $\tau_{n\psi}$ acting on the fracture plane.
2. If σ_n is a tensile stress it promotes fracture together with the shear stress $\tau_{n\psi}$ or even alone for $\tau_{n\psi} = 0$.
   In contrast to that σ_n impedes fracture if it is a compressive stress by raising the fracture resistance of the fracture plane against shear fracture with increasing compressive stress σ_n.

Experimental experience has shown, that transverse shear stressing $\tau_{\perp\perp}$ causes a fracture not in the action plane of $\tau_{\perp\perp}$ but in the plane of the tensile principal stress which has the same magnitude as $\tau_{\perp\perp}$. Therefore, a rule has been formulated by Puck for the case of combined stresses σ_2, σ_3, τ_{23} or σ_{II}, σ_{III} respectively:

If the UD-lamina is just stressed in the plane of transverse isotropy (only stresses σ_2, σ_3, τ_{23}) fracture occurs either as tensile fracture due to σ_n as the highest normal stress or as shear fracture due to τ_{nt} – impeded by a transverse compressive stress σ_n. Which of the two kinds occurs depends on the ratios of the stresses σ_2, σ_3, τ_{23}.

This additional rule might not be valid for composites with an unusually small ratio between transverse compressive and transverse tensile strength $R_\perp^c/R_\perp^t < 2$. Experimental results for thermoplastic FRP (TP-FRP) like PEEK (Polyetheretherketon) show however, that for those thermoplastic materials the rule is valid just as for thermoset FRP[18] [Kuhnel 2008].

---

[18] These investigations show, too, that for PEEK the ratio $R_\perp^c/R_\perp^t$ is larger than 2,5.

If there is a UD-material with an extraordinary low ratio $R_\perp^c/R_\perp^t$ it is expected that the rule just mentioned is not fulfilled. $\tau_{\perp\perp}$-stressing would instead lead to a mixed mode fracture of combined $\sigma_\perp^t$-stressing and $\tau_{\perp\perp}$-stressing.

4.2.4 Fracture resistance of the action plane

In almost all failure criteria the used mathematical functions are calibrated by using the maximum sustainable stresses for uniaxial tensile or compressive stresses and pure shear stresses. The absolute values of these stresses are given by the so called "strengths" R. For stress based fracture criteria of UD-composites the so called basic strengths

$R_\parallel^t$, $R_\parallel^c$, $R_\perp^t$, $R_\perp^c$, $R_{\perp\perp}$, $R_{\perp\parallel}$

are used for calibration. All these are given as positive values, also the compressive strengths. This is a source of misunderstanding and mistakes, because in fracture criteria sustainable stresses have to be used. As compressive stresses have a negative sign, the corresponding maximum sustainable uniaxial stresses are for a UD-composite "$-R_\perp^c$" and "$-R_\parallel^c$".

The basic strengths correspond to the fracture limit under pure $\sigma_\parallel$, $\sigma_\perp$, $\tau_{\perp\perp}$ or $\tau_{\perp\parallel}$ stressing, respectively. These fracture limits are measured without giving any attention to the kind of failure (failure by yielding, brittle fracture, crushing etc.). Thus, $R_{\perp\perp}$ is defined as the transverse shear stressing $\tau_{\perp\perp}$ leading to failure in the absence of all other stresses. However, experimental experience shows that in this case in brittle material fracture takes place on a plane inclined by 45° to the action plane of the applied $\tau_{\perp\perp}$ (compare Fig. 26). On this plane pure transverse tension $\sigma_\perp^t$ is acting. This is not considered when conventional failure criteria are used. The shear strength $R_{\perp\perp}$ is calculated by dividing the applied shear load by the area it has been applied on.

In the same way as $R_{\perp\perp}$, the transverse compressive strength $R_\perp^c$ corresponds to $\sigma_\perp^c$ stressing leading to failure in the absence of all other stresses. However, it has been proved that by no means fracture occurs on the action plane of the $\sigma_\perp^c$-stressing but on a plane inclined by some ±54° (see above). On this plane transverse shear stressing $\tau_{\perp\perp}$ is dominant and responsible for IFF. But this experience is also not considered when the transverse compressive strength $R_\perp^c$ has to be determined. $R_\perp^c$ is also found by dividing the applied transverse compressive failure load by the area of the cross-section of the specimen.

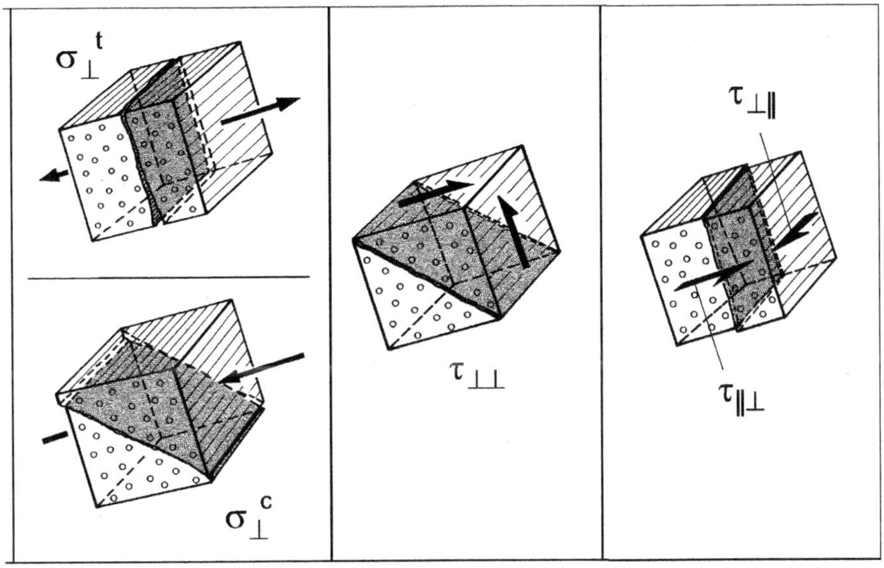

Fig. 26. Stressings of the UD-lamina causing IFF and corresponding fracture planes

As long as a failure criterion is formulated with stresses defined in the natural coordinate system of the layer (x_1, x_2, x_3) the basic strengths are the usual calibration parameters. Puck's IFF-criteria, however, are formulated with the stresses of the action plane (σ_n, τ_{nt}, τ_{n1}). Consequently, the corresponding fracture resistances of the action plane (defined as R^A in order to distinguish them from the basic strengths R) are needed.

Instead of employing the standard form of a failure condition

$$F(\sigma_1, \sigma_2, \sigma_3, \tau_{23}, \tau_{31}, \tau_{21}, R_\parallel^t, R_\parallel^c, R_\perp^t, R_\perp^c, R_{\perp\perp}, R_{\perp\parallel}) = 1 \quad \text{(Eq. 22)}$$

a IFF-fracture condition is formulated in the form

$$F\left(\sigma_n(\theta_{fp}), \tau_{nt}(\theta_{fp}), \tau_{n1}(\theta_{fp}), R_\perp^{At}, R_{\perp\perp}^A, R_{\perp\parallel}^A\right) = 1. \quad \text{(Eq. 23)}$$

Looking on (Eq. 23) it is important to remember that the stresses $\sigma_n(\theta)$, $\tau_{nt}(\theta)$, $\tau_{n1}(\theta)$ are the three stresses which in the most general case are acting simultaneously on one and the same action plane[19]. The action plane which is inclined by $\theta = \theta_{fp}$ is the fracture plane because on this action plane the risk of fracture is the highest one.

[19] That σ_n, τ_{nt}, τ_{n1} have the same action plane can be seen from the index "n" of all three stresses.

If all three stresses σ_n, τ_{nt}, τ_{n1} act simultaneously on an action plane, they represent a combined stressing consisting of a $\sigma_\perp$-stressing, a $\tau_{\perp\perp}$-stressing and a $\tau_{\perp\|}$-stressing.

In order to answer the question whether a certain combination of σ_n, τ_{nt}, τ_{n1} can be sustained by a fiber parallel action plane a fracture criterion has to be formulated mathematically. On the one hand it should contain the stresses σ_n, τ_{nt}, τ_{n1} acting simultaneously on a common action plane. On the other hand a fracture criterion of this kind can only be calibrated by experimental values for the sustainable stress of the action plane, when a $\sigma_\perp$-stressing is acting alone, when a $\tau_{\perp\perp}$-stressing is acting alone and when a $\tau_{\perp\|}$-stressing is acting alone. This means that fracture tests have to be performed with uniaxial $\sigma_\perp$, pure $\tau_{\perp\perp}$ and pure $\tau_{\perp\|}$.

If in such a test with – for instance – a single tensile stressing $\sigma_\perp^t$ the fracture would occur in the action plane of the applied stressing this would be the sustainable tensile stressing of the action plane. Puck has introduced the following expression:

- "*Fracture resistance of the action plane*" and the corresponding symbol R^A.

The definition of R^A is the following:

A fracture resistance of the action plane is the resistance (expressed in the dimension of a stress) by which an action plane resists its own fracture caused by a single stressing ($\sigma_\perp$ or $\tau_{\perp\perp}$ or $\tau_{\perp\|}$) acting in the action plane under consideration.

This definition shows how important it is for the fracture experiments to check whether or not the applied $\sigma_\perp$- or $\tau_{\perp\perp}$- or $\tau_{\perp\|}$-stressing really leads to a fracture in the action plane of the applied stress.

Some of the fracture resistances of the action plane used in (Eq. 23) are identical to the basic strengths as will be shown in the following. However, it is important to understand that the underlying question is different. The question is for instance

> "To which value $\sigma_\perp^t$ has to be increased in order to provoke failure **on its** action plane?"

and not

> "To which value $\sigma_\perp^t$ has to be increased in order to provoke fracture (on any plane)?"

There are three single fracture resistances to be differentiated:

$R_{\perp}^{At}$ = Resistance of the action plane against its fracture due to transverse tensile stressing $\sigma_{\perp}^{t}$ acting in that plane.
$R_{\perp\perp}^{A}$ = Resistance of the action plane against its fracture due to transverse shear stressing $\tau_{\perp\perp}$ acting in that plane.
$R_{\perp\parallel}^{A}$ = Resistance of the action plane against its fracture due to longitudinal shear stressing $\tau_{\perp\parallel}$ acting in that plane.

In the following some examples will be given to illustrate the meaning of action plane fracture resistances. The first example is a pure longitudinal shear stressing $\tau_{\perp\parallel}$ caused by $\tau_{21} \neq 0$ in the absence of all other stresses. This will obviously occur in a test with a pure longitudinal shear stressing realized by applying – for instance – a pure shear stress τ_{21} in absence of all other stresses. In this case a shear stress $\tau_{nl} = \tau_{21} \cdot \cos\theta$ is acting on a plane inclined by the angle θ to the x_2-plane on which τ_{21} is acting. Obviously, for all angles $\theta \neq 0°$ the shear stress τ_{nl} is smaller than τ_{21}. Fracture must therefore occur on the action plane of τ_{21} ($\theta_{fp} = 0°$) and therefore

$$R_{\perp\parallel}^{A} = R_{\perp\parallel} \qquad \text{(Eq. 24)}$$

This means that the action plane fracture resistance $R_{\perp\parallel}^{A}$ is identical to the longitudinal shear strength $R_{\perp\parallel}$.

Transverse tension – for instance σ_2^{t} – normally leads in the absence of other stresses to fracture in its action plane, too, and therefore:

$$R_{\perp}^{A\,t} = R_{\perp}^{t} \qquad \text{(Eq. 25)}$$

This, however, is not as evident as in the case of pure longitudinal shear. In the case of transverse tension σ_2^{t} there is a combination of transverse tension ($\sigma_{\perp}^{t} = \sigma_2 \cos^2\theta$) and transverse shear ($\tau_{\perp\perp} = -0.5\ \sigma_2 \sin 2\theta$) on planes $\theta \neq 0$. If the resistance of the action plane against $\tau_{\perp\perp}$ shear fracture would be considerably smaller than that against transverse tension a different fracture angle $\theta_{fp} \neq 0°$ would occur.

The third relevant fracture resistance $R_{\perp\perp}^{A}$ cannot as easily be determined as $R_{\perp\parallel}$ and $R_{\perp}^{t}$. When applying a pure τ_{23} stress to a unidirectional layer in the absence of any other stress there are three planes on which just one stressing is acting:

- $\theta = 0°$: action plane of τ_{23} ($\tau_{\perp\perp}$-stressing)
- $\theta = 90°$: action plane of τ_{32} ($\tau_{\perp\perp}$-stressing)
- $\theta = 45°$: action plane of the principal tensile stress ($\sigma_{\perp}^{t}$-stressing) resulting from τ_{23} and being of same magnitude as τ_{23} and τ_{32}.

According to Puck's hypotheses fracture occurs on the plane with the highest stress exposure. Fracture tests on UD-specimens loaded by $\tau_{\perp\perp}$ show fracture under $\theta_{fp} = 45°$. This leads to the clear conclusion that the fracture resistance $R_{\perp}^{At}$ against transverse tension is smaller than the fracture resistance $R_{\perp\perp}^{A}$ against transverse shear. In other words: With a transverse shear test not $R_{\perp\perp}^{A}$ but instead the resistance against transverse tension $R_{\perp}^{At}$ is measured.

In fact there is no way to measure $R_{\perp\perp}^{A}$ for intrinsically brittle fiber reinforced plastics directly. However, the unidirectional compression test offers an indirect way for the determination of $R_{\perp\perp}^{A}$ as long as the hypothesis for pure transverse stressing (compare chapter "Fracture hypotheses") is valid. If just σ_2^{c} is applied, a fracture angle of $\theta_{fp} \approx \pm 54°$ results. On this fracture plane a combination of transverse shear $\tau_{\perp\perp}$ and transverse compression $\sigma_{\perp}^{c}$ is acting. According to fracture hypothesis 2, a transverse compressive stress adds an additional fracture resistance to the "intrinsical" fracture resistance $R_{\perp\perp}^{A}$. This additional fracture resistance can be calculated by the used mathematical fracture condition. In this way the test result can be corrected in order to find $R_{\perp\perp}^{A}$. Thus, the calculated value of $R_{\perp\perp}^{A}$ depends on the fracture model used. Therefore this $R_{\perp\perp}^{A}$ is perhaps not the "real" value of the material property $R_{\perp\perp}^{A}$ but it can be used for the fracture model by which it has been calculated. At least it is guaranteed that by using this same fracture model in a fracture analysis for the special case of uniaxial compression the result will be $(\sigma_2)_{fr} = -R_{\perp}^{c}$ with $R_{\perp}^{c}$ being the real transverse compressive strength.

The example of longitudinal shear is well suited to explain again the difference between a fracture resistance (of the action plane) and a conventional basic strength. The stress τ_{21} is acting on the x_2-plane parallel to the fibers and the corresponding shear stress τ_{12} of the same magnitude is acting on the x_1-plane. This shear stress on the x_1-plane is correctly addressed as τ_{12}. However, there is just one strength parameter $R_{\perp\parallel}$ and it does not contain the information about the fracture plane. This could be either the x_2- or the x_1-plane or even another plane. The corresponding fracture resistances of the two planes under consideration differ, however, considerably. Namely, the fracture resistance ($R_{\perp\parallel}^{A}$) of the x_2-plane – parallel to the fibers – is much smaller than the fracture resistance $R_{\parallel\perp}^{A}$ of the x_1-plane perpendicular to the fibers. This phenomenon can easily be understood: On the x_2-plane the stressing $\tau_{\perp\parallel}$ provokes Inter Fiber Fracture. The fracture plane runs parallel to the fibers whereas a fracture on the x_1-plane would be a Fiber Fracture (shear off of the fibers)!

Obviously it makes no sense to talk about a resistance of the action plane against fracture due to transverse compressive stressing $\sigma_\perp^c$ because a compressive stress cannot produce a fracture in its own action plane. Summing up this, there are three action plane fracture resistances to be differentiated:

$R_\perp^{At}$ = Resistance of the action plane against its fracture due to transverse tensile stressing $\sigma_\perp^t$ acting in that plane. For structural UD-material under normal temperature and humidity conditions it can be expected that $R_\perp^{At} = R_\perp^t$ is valid,

$R_{\perp\perp}^A$ = Resistance of the action plane against its fracture due to transverse shear stressing $\tau_{\perp\perp}$ acting in that plane. Attention has to be paid to the fact that $R_{\perp\perp}^A \neq R_{\perp\perp}$,

$R_{\perp\parallel}^A$ = Resistance of the action plane against its fracture due to longitudinal shear stressing $\tau_{\perp\parallel}$ acting in that plane. For elementary reasons $R_{\perp\parallel}^A = R_{\perp\parallel}$.

Experience has shown that most mechanical engineers are not familiar with the difference between "strength" and "action plane fracture resistance". It could be helpful to realize the different underlying question. Dealing with strengths the underlying question is

"To which value can a certain stressing (for instance uniaxial stressing $\sigma_\perp^t$) be increased until failure occurs somewhere (and somehow) in the evenly stressed volume of the test specimen?"

Dealing with action plane fracture resistances the underlying question is:

> "To which value can a certain stressing be increased until a fracture occurs on the action plane in which the evenly applied stressing is acting?"

4.2.5 Visualization of the stress/strength problem

Preliminary remark:
The content of the following chapter is not required to be read by somebody who is interested in the application only of the algorithms of Puck's laminate failure analysis. However, it is strongly recommended to have a look on the astonishing possibilities of the visualization because they present a much better understanding of the results of the mathematical treatment of the physical problems which is explained in the subsequent chapters. But the reader being in a hurry may continue with the chapter "Universal 3-D-formulation of the action plane related IFF-criteria".

4.2.5.1 Mohr's Circles and Fracture Envelopes

Within a failure analysis independently of the kind of material the engineer has to complete two tasks: First the stresses of the material caused by the loads have to be calculated and in a second step these stresses have to be compared to the failure limit. If the material fails by brittle fracture, Mohr's circles are helpful for the first task and Mohr's envelope for the second one.

Because of the fact that Puck's IFF hypotheses are based on the ideas of Mohr [Mohr 1900] and Coulomb [Coulomb 1776] one can partially use these classical solutions. However, it must be recognized that the approaches of Mohr and Coulomb are valid only for macroscopically isotropic materials. In the isotropic case fracture due to an arbitrary three dimensional state of stress can be treated as a 2D-problem. Any stress combination leading to fracture can be visualized by means of the well known Mohr's circle and Mohr's envelope.

The reason for this opportunity of simplification is the isotropy and the validity of Mohr's fracture hypothesis for brittle fracture. Any 3D-stress combination (σ_1, σ_2, σ_3, τ_{23}, τ_{31}, τ_{21})[20] – to use the coordinates which are established for the UD-lamina (compare Fig. 16) – can be transformed to an equivalent (σ_a, σ_b, σ_c) stress state without any shear stresses on the action planes of σ_a, σ_b, σ_c where σ_a, σ_b and σ_c are the so called principal stresses. If $\sigma_b > \sigma_a$ by $\sigma_b < \sigma_a$ being the intermediate principal stress, fracture depends according to Mohr only on the major and minor principal stresses σ_b and σ_c (Fig. 27). In the (σ, τ)-diagram Fig. 27 points on the circumference of a circle with the coordinates σ and τ present the stresses σ and τ on section planes through the material, the normal of which is perpendicular to one of the three axes a, b or c, that means perpendicular to one of the three principal stresses. It can be shown theoretically that the stresses σ and τ of all other section planes are given by points located in the hatched areas of Fig. 27.

Therefore the highest shear and normal stresses occurring in the material under the given stresses depend only on the two extreme principal stresses. In Fig. 27 σ_b and σ_c are chosen as the extreme principal stresses for demonstration. However, depending on the load case it could be any other pair of the three principal stresses σ_a, σ_b, σ_c too. The axes a, b, c of the Cartesian coordinate system in which the acting stress combination can be presented for the isotropic material *changes its spatial orientation* depending on the actual stress state (σ_1, σ_2, σ_3, τ_{23}, τ_{31}, τ_{21}). That means that gen-

[20] The additional corresponding shear stresses follow from $\tau_{23}=\tau_{32}$, $\tau_{31}=\tau_{13}$, $\tau_{21}=\tau_{12}$.

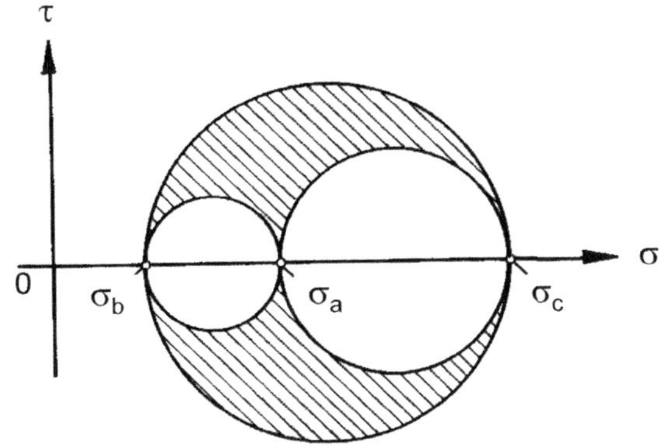

Fig. 27. System of the three Mohr-circles for isotropic material

erally the directions a, b, c in which the principal stresses σ_a, σ_b, σ_c are acting do not coincide with any of the chosen axes x_1, x_2, x_3 of the original reference coordinate system for the actual stress combination (σ_1, σ_2, σ_3, τ_{23}, τ_{31}, τ_{21}), see Fig. 28.

If there is isotropy there is no difficulty in working with a coordinate system with axes a, b, c of the principal stresses σ_a, σ_b, σ_c which for different states of stress has a different spatial orientation and there is also no problem to work with an unlimited number of coordinate systems for the axes of the principal stresses. However, obviously such versatility does not exist for transversely-isotropic materials like the UD-lamina where the fixed direction of the fibers prevents such a procedure. When dealing with transversely-isotropic material a change of the direction of coordinate axes

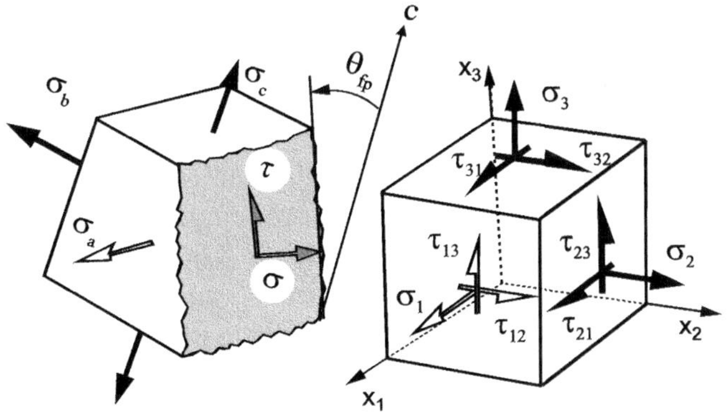

Fig. 28. The stresses σ and τ given by the coordinates of a point on the largest circle in Fig. 27 are acting on a potential fracture plane which is parallel to σ_a. The direction of σ and τ is perpendicular to σ_a

is at least possible in the plane of transverse isotropy which is perpendicular to the fixed axis parallel to the fibers.

4.2.5.2 Mohr's circle and Mohr's envelop for a transversely-isotropic UD-lamina under plane (σ_2, σ_3, τ_{23})-stress

In a UD-lamina the stresses σ_2, σ_3, τ_{23} which are acting on action planes being parallel to the fixed fiber direction x_1 are the only stresses to which Mohr's circle can be applied effectively. Transverse isotropy allows to switch over from the (x_2, x_3)-coordinate system to a (x_n, x_t)–coordinate system.

In order to get familiar with the derivation and application of Mohr's circle, a plane (σ_2, σ_3, τ_{23}) stress state will be investigated now. The stresses acting on the plane which is inclined by an angle θ to the x_3-axis (compare Fig. 33) can be determined by (Eq. 1) and (Eq. 2). Using the well known theorems for "sin" and "cos" (Eq. 1) and (Eq. 2) gives the form

$$\sigma_n(\theta)=\frac{1}{2}(\sigma_2+\sigma_3)+\frac{1}{2}(\sigma_2-\sigma_3)\cdot\cos 2\theta+\tau_{23}\cdot\sin 2\theta \qquad \text{(Eq. 26)}$$

$$\tau_{nt}(\theta)=-\frac{1}{2}(\sigma_2-\sigma_3)\cdot\sin 2\theta+\tau_{23}\cdot\cos 2\theta \qquad \text{(Eq. 27)}$$

These equations look already like the mathematical description of a circle using the angle 2θ as a parameter. This can be confirmed in the following way: Rearrangement of terms, squaring of (Eq. 26) and adding up of the squared form of (Eq. 27) finally leads to the following expression:

$$\left[\sigma_n(\theta)-\frac{1}{2}(\sigma_2+\sigma_3)\right]^2+\left[\tau_{nt}(\theta)\right]^2=\left[\frac{1}{2}(\sigma_2-\sigma_3)\right]^2+\tau_{23}^{\,2} \qquad \text{(Eq. 28)}$$

(Eq. 28) confirms the very welcome opportunity for the visualization of stress transformation because (Eq. 28) is the mathematical description of a circle with the center on the σ_n-axis at $0.5(\sigma_2+\sigma_3)$ and the radius $\sqrt{\left[\frac{1}{2}(\sigma_2-\sigma_3)\right]^2+\tau_{23}^{\,2}}$. This circle is illustrated in Fig. 29.

Figure 29 presents a graphical procedure by which the stresses σ_{II} and σ_{III} can be determined if σ_2, σ_3 and τ_{23} are known.

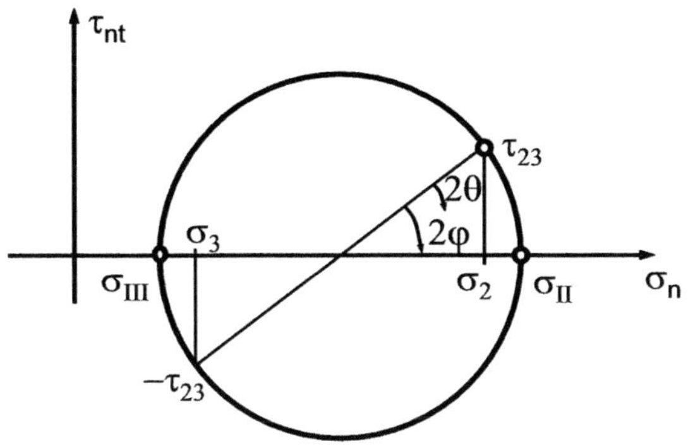

Fig. 29. Mohr circle for $\{\sigma_2, \sigma_3, \tau_{23}\}$

On planes characterized by the angles φ and (φ+90°) there is no shear stress τ_{nt}. The corresponding normal stresses $\sigma_n(\varphi)$ and $\sigma_n(\varphi+90°)$ are designated by σ_{II} and σ_{III} and are called "extreme normal stresses of the transversely-isotropic plane". In general they are not real principal stresses. This is explained later.

It is important to realize that for instance an angle φ in the real material with a mathematically positive sense of rotation corresponds to an angle of 2φ with a mathematically negative sense of rotation in the Mohr-circle!

The "extreme normal stresses σ_{II} and σ_{III} of the transversely isotropic plane" and the angle φ (compare Fig. 29) can also very easily be found in the usual analytical way by setting $\tau_{nt} = 0$ in (Eq. 27). It is

$$2\varphi = \arctan\frac{2\tau_{23}}{\sigma_2 - \sigma_3} \qquad \text{(Eq. 29)}$$

and with (Eq. 26)

$$\sigma_{II} = \frac{1}{2}(\sigma_2 + \sigma_3) + \frac{1}{2}(\sigma_2 - \sigma_3)\cos 2\varphi + \tau_{23}\sin 2\varphi \qquad \text{(Eq. 30)}$$

$$\sigma_{III} = \frac{1}{2}(\sigma_2 + \sigma_3) - \frac{1}{2}(\sigma_2 - \sigma_3)\cos 2\varphi - \tau_{23}\sin 2\varphi\,. \qquad \text{(Eq. 31)}$$

With known extreme normal stresses σ_{II} and σ_{III} the stresses $\sigma_n(\Theta)$ and $\tau_n(\Theta)$ on planes inclined to the direction of σ_{II} by the angle Θ can be calculated with (Eq. 32) and (Eq. 33). These equations correspond to (Eq. 26) and (Eq. 27).

$$\sigma_n(\Theta) = \frac{1}{2}(\sigma_{II} + \sigma_{III}) + \frac{1}{2}(\sigma_{II} - \sigma_{III})\cos 2\Theta \qquad \text{(Eq. 32)}$$

$$\tau_{nt}(\Theta) = -\frac{1}{2}(\sigma_{II} - \sigma_{III})\sin 2\Theta \qquad \text{(Eq. 33)}$$

Mohr's circle as shown in Fig. 30 is based on (Eq. 32) and (Eq. 33). The coordinates σ_n and τ_n of a point P at a radian distance of 2Θ from σ_{II} are the stresses acting on a plane inclined by Θ to the action plane of σ_{II} (compare Fig. 30). The centre of the circle has a distance of $0.5(\sigma_{II}+\sigma_{III}) = 0.5(\sigma_2+\sigma_3)$ from the point of origin. The radius of the circle is $0.5(\sigma_{II}-\sigma_{III})$. The term $(\sigma_{II}-\sigma_{III})$ can be negative or positive. To make sure that the graphical construction is in line with (Eq. 32) and (Eq. 33) (positive radius), the sense of rotation and the starting point of the Θ-measurement has to be adapted in a way that $\sigma_n(\Theta)$ and $\tau_n(\Theta)$ are correct by means of magnitude and algebraic sign, see sketch on the bottom of Fig. 30.

All Mohr's circles for arbitrary states of stress have their centers on the σ_n – axis. Each point on the circumference of such a circle with coordinates σ_n and τ_{nt} belongs to a certain section through the material with an angle Θ (in reality). The stress vector belonging to such a section has its starting point in the origin of the (σ_n, τ_{nt})-coordinate system and its tip touches the point $(\sigma_n(\Theta), \tau_{nt}(\Theta))$ on the circumference of Mohr's circle.

No vector tip belonging to a vector representing a sustainable state of stress can exceed a certain fracture limit of the material. Therefore Mohr

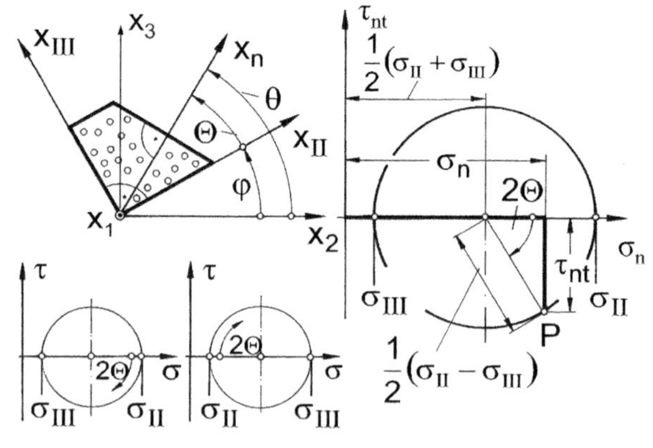

Fig. 30. Mohr circle and derivation of stresses on inclined planes

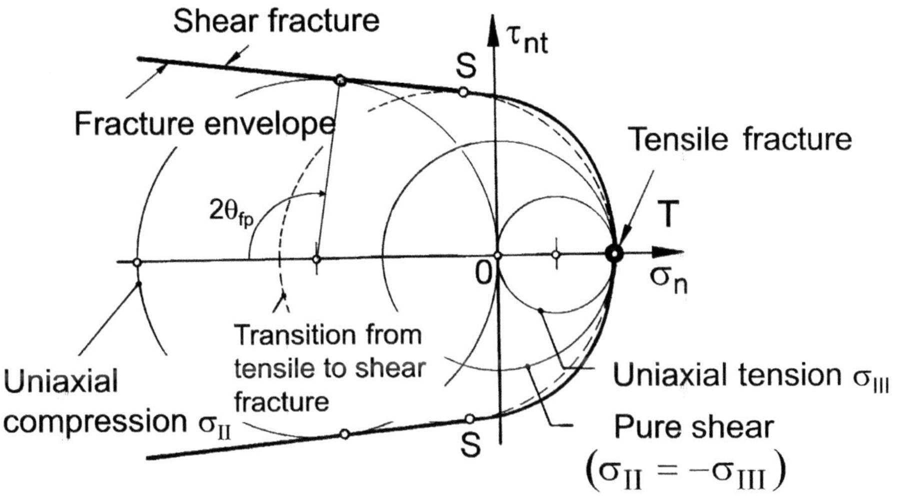

Fig. 31. Mohr's circles for special stress states and Mohr's fracture envelope

has not only provided the helpful visualization of stresses acting on inclined action planes by Mohr's circle but he has also tried to draw a fracture limit of the material as a curve in the (σ_n, τ_{nt})-diagram. This is called "Mohr's envelope", see Fig. 31. It envelops all sustainable stress combinations. Obviously no Mohr's circle for sustainable stress can exceed this envelope; it can at the most touch it.

It is important to realize that if the contact point of a Mohr's circle with the Mohr-envelope is found, not only the combined stresses σ_n and τ_{nt} on the fracture plane but as well the fracture angle θ_{fp} can be found in the diagram, see for instance Fig. 31.

For a UD-lamina the fracture envelope is chosen by Puck as a parabola in the area of $\sigma_n < 0$ and an ellipse in the area of $\sigma_n > 0$. At $\sigma_n = 0$ the curve might have a sharp bend or – more mathematically spoken – a discontinuity in terms of gradient.

According to the fracture hypotheses discussed in detail in the chapter "Fracture hypotheses", there is (as far as IFF is concerned) either tensile fracture or shear fracture in a UD-lamina stressed by (σ_2, σ_3, τ_{23}, 0, 0). Consequently, on the fracture plane there is either a $\sigma_n > 0$ (pure $\sigma_\perp^t$) stressing or a combination of $\sigma_n < 0$ and τ_{nt} (($\sigma_\perp^c$, $\tau_{\perp\perp}$)-stressing). A combined ($\sigma_\perp^t$, $\tau_{\perp\perp}$)-stressing or a pure $\tau_{\perp\perp}$-stressing on the fracture plane are both not possible according to the additional "rule" given with the two fracture hypotheses (compare chapter "Fracture hypotheses").

Under these conditions, all Mohr circles describing states of stress which lead to tensile fracture touch the fracture envelope at the same point,

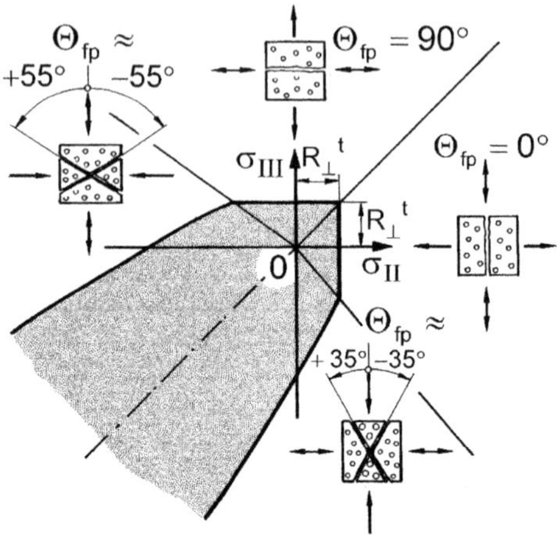

Fig. 32. (σ_{II}, σ_{III}) fracture curves resulting from Mohr's fracture envelope

namely T($R_\perp^t$, 0) (compare Fig. 31). In a (σ_{II}, σ_{III})-diagram the corresponding fracture curve for tensile fracture consists of two straight lines, running at a distance of $R_\perp^t$ parallel to the σ_{II}- and σ_{III}-axes (compare Fig. 32). This corresponds to Paul's [Paul 1961] findings for intrinsically brittle materials. In the first quadrant of the (σ_{II}, σ_{III})-diagram tensile fracture occurs as a consequence of σ_{II} ($\theta_{fp} = 0°$) – in the case of $\sigma_{II} > \sigma_{III}$ – or as a consequence of σ_{III} ($\theta_{fp} \pm 90°$) in the case of $\sigma_{III} > \sigma_{II}$. For the straight fracture lines $\sigma_{II} = R_\perp^t$ and $\sigma_{III} = R_\perp^t$ Paul introduced the term "tensile cut offs". In the 2nd and 4th quadrant, either tensile or shear fracture can occur. What fracture occurs depends on the ratio σ_{II}/σ_{III}. In the 3rd quadrant, where both σ_{II} and σ_{III} are compressive stresses fracture is always a shear fracture. However, the fracture plane is not the plane with the highest shear stressing ($\theta_{fp} = \pm 45°$ with $\tau_{nt} = 0.5|(\sigma_{II}\text{-}\sigma_{III})|$; in Mohr's circle straight above and below the centre point at $\pm 2\theta_{fp} = \pm 90°$). Instead, the fracture angle deviates by $\approx \pm 5°$ to $\pm 10°$ from the plane with the highest shear stress. This becomes understandable when regarding the normal stress σ_n^c acting simultaneously to τ_{nt}. This compressive normal stress is reduced considerably when the angle deviates a bit from $\pm 45°$, whereas the shear stress τ_{nt} stays nearly the same (compare Fig. 25). Knowing that compressive normal stress impedes shear fracture, this explains why the observed angle of shear fracture is not $\pm 45°$ but somewhat larger. This has been experimentally confirmed with uniaxial compression tests, see Fig. 8.

Figure 31 shows that the fracture envelope and the Mohr's circles have no points of contact between the points T ($R_\perp^t$, 0) and S (point of transition from tensile fracture to shear fracture). The dashed circle in Fig. 31 marks the "circle of transition" from tensile to shear fracture. For this special (σ_{II}, σ_{III})-compression/tension stress combination there can either occur a shear fracture or a tensile fracture. Also the circle for pure transverse shear shows contact only in T, that means $R_{\perp\perp} = R_{\perp}^{t}$ is to be expected.

The course of the fracture envelope between the points T and S does not seem to be relevant for the states of stress treated so far. There are no contact points on it. It is a so called "dead branch" of the fracture envelope. However, from the following paragraph it will become clear, that the course of the (σ_n, τ_{nt}) fracture curve between T and S has an influence on the fracture envelope for 3D-stressings (σ_n, τ_{nt}, τ_{n1}), the so called Master Fracture Body (MFB). Such a 3D-stressing is present as soon as there is longitudinal shear $\tau_{\varpi 1}$, that means τ_{21} and/or τ_{31} too.

4.2.5.3 From Mohr's circle to Puck's cosine-shaped cylinder

The idea of Mohr's circles and Mohr's envelope is used now as the basis for a corresponding procedure for transversely-isotropic materials with the most general 3D-state of stress. Figure 33 shows all stresses and their correct designation of a UD-lamina. The stress σ_1 leads to a fiber parallel stressing $\sigma_{\|}$, whereas both the stress σ_2 and the stress σ_3 cause a transverse stressing $\sigma_{\perp}$. The stress $\tau_{23} = \tau_{32}$ leads to a transverse shear stressing $\tau_{\perp\perp}$ and τ_{21} and τ_{31} both lead to a longitudinal shear stressing $\tau_{\perp\|}$. That means that obviously all three fracture resistances $R_{\perp}^{At}$, $R_{\perp\perp}^{A}$, $R_{\perp\|}^{A}$ of the fiber parallel action plane will now generally influence the IFF-process.

In order to get to visualization also of the most general stress situation one must try not to use more than three variables. For this reason one should combine as much stresses as possible in "resulting" stresses. The

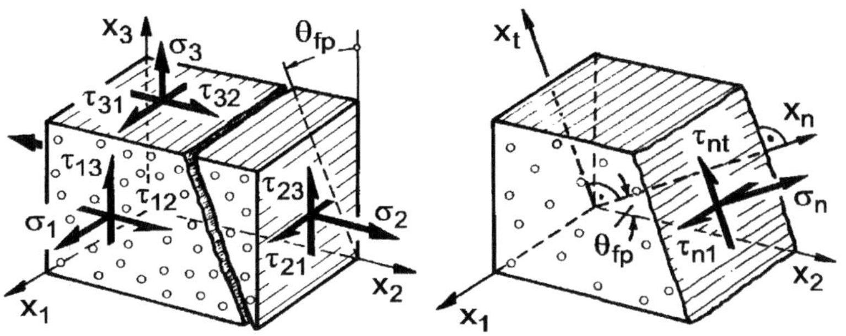

Fig. 33. Stresses of the UD-lamina and stresses on a possible (IFF)-fracture plane which is parallel to the fibers

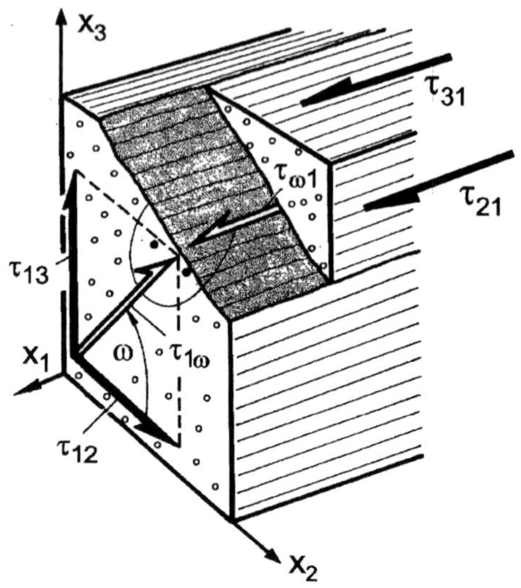

Fig. 34. Uniting τ_{12} and τ_{13} to $\tau_{1\varpi}$ and τ_{21} and τ_{31} to $\tau_{\varpi 1}$

longitudinal shear stresses τ_{12} and τ_{13} and their corresponding stresses τ_{21} and τ_{31} are quite special in this respect. Figure 34 shows that τ_{12} and τ_{13} are in fact components of only one longitudinal shear stress $\tau_{1\varpi}$. The two components are $\tau_{12} = \tau_{1\varpi}\cos\varpi$ and $\tau_{13} = \tau_{1\varpi}\sin\varpi$. Adequate equations can also be written with the corresponding stresses τ_{21}, τ_{31} and $\tau_{\varpi 1}$ (this is the stress provoking IFF in its fiber parallel action plane). Implicitly the fact that the longitudinal shear stresses can easily be united in one stress means: there does exist only one relevant longitudinal shear stress and this is $\tau_{\varpi 1}$. As long as only $\tau_{\omega 1}$ acts in the absence of any other stress, IFF will occur in the plane in which the resulting longitudinal shear stress $\tau_{\varpi 1}$ is acting as soon as $\tau_{\varpi 1} = R_{\perp\parallel}$ is reached. The fracture plane is orientated perpendicular to the direction of $\tau_{1\varpi}$ as shown in Fig. 34.

It is important to note that the stresses σ_{II} and σ_{III} are only true principal stresses like σ_b and σ_c respectively, if there is no longitudinal shear $\tau_{\varpi 1}$. In this special case σ_I, σ_{II}, σ_{III} are real principal stresses. As soon as $\tau_{\varpi 1} \neq 0$, there is longitudinal shear on the action plane of σ_{II} and/or σ_{III}. The action plane of "true" principal stresses is however free of any shear stresses.

Obviously the combined action by which the stresses σ_{II}, σ_{III} and $\tau_{\varpi 1}$ cause an IFF will depend on the direction of the shear stress $\tau_{\varpi 1}$ compared to the directions of the normal stresses σ_{II} and σ_{III}. But due to the transverse isotropy of the UD-lamina (with respect to the x_1-axis) not the absolute

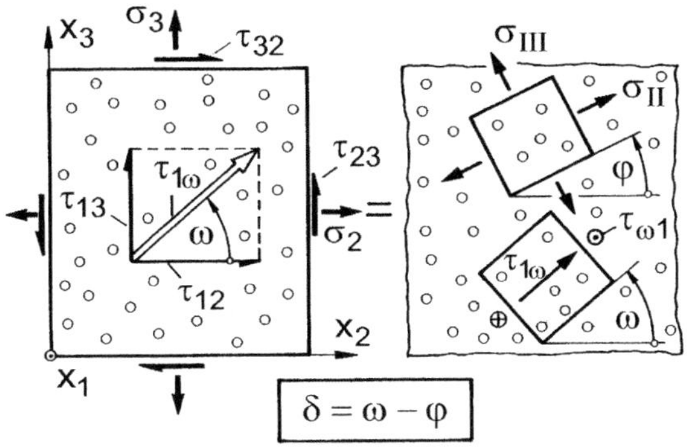

Fig. 35. Combining σ_2, σ_3, τ_{23} to σ_{II} and σ_{III} and τ_{12} and τ_{13} to $\tau_{1\varpi}$; definition of angle δ

values of the angles ϖ and φ are relevant for the risk of fracture. Instead, the difference δ between ϖ and φ is relevant (see Fig. 35):

$$\delta = \omega - \varphi . \qquad \text{(Eq. 34)}$$

Because of the meaning of ω and φ (see Fig. 35 and (Eq. 29) δ can be written as:

$$\delta = \arctan\frac{\tau_{31}}{\tau_{21}} - \frac{1}{2}\arctan\frac{2\tau_{23}}{\sigma_2 - \sigma_3} . \qquad \text{(Eq. 35)}$$

Longitudinal shear stress $\tau_{\varpi 1}$ cannot be eliminated like a transverse shear stress τ_{23} by the introduction of stresses like σ_{II} and σ_{III}. Thus, in the general case fracture has to be calculated for a (σ_1, σ_{II}, σ_{III}, 0, $\tau_{\omega 1}$) state of stress.

For the moment it is assumed that just as in the case of isotropic material (with σ_a being the intermediate principal stress) the fiber parallel stress σ_1 has no influence on the fracture angle Θ_{fp} and no influence on the magnitude of the fracture stresses σ_{II}, σ_{III}, $\tau_{\omega 1}$ at IFF – this is at least correct as long as σ_1 is considerably smaller than $\approx$ 50 % of the fiber fracture strength $|R_{||}|$.

In the particular case that the longitudinal shear $\tau_{\varpi 1}$ acts alone ($\sigma_{II} = \sigma_{III} = 0$), fracture occurs in the action plane of $\tau_{\varpi 1}$ which is turned by the angle difference $\delta = \varpi$-φ with respect to the action plane of σ_{II} (compare Fig. 35). In a further special case, $\tau_{\varpi 1}$ acts on the fracture plane of the tensile fracture provoked by (σ_{II}, σ_{III}) alone. In this case, the longitudinal shear simply “helps” provoking fracture without changing the fracture angle. The

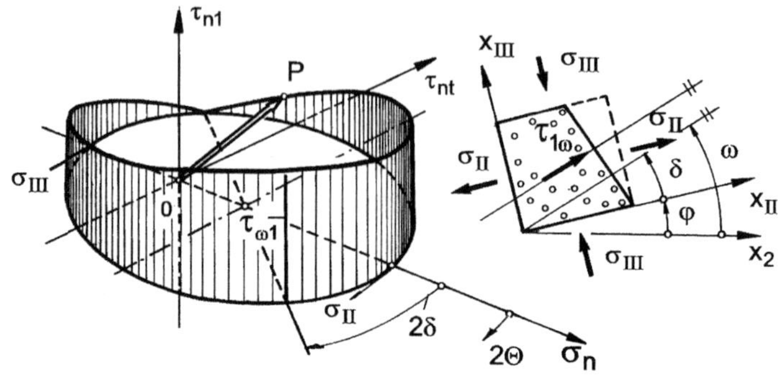

Fig. 36. Mohr's circle with longitudinal shear τ_{n1} representation for UD-lamina

common presence of σ_{II}, σ_{III} and $\tau_{\varpi 1}$ leads to a combined ($\sigma_{\perp}^{t}$, 0, $\tau_{\perp\parallel}$) mode of fracture, that means a Mode A-fracture.

The tensile fracture occurs on the action plane of σ_{II} or that of σ_{III} (depending on which of both stresses is larger). If, however, the action plane of $\tau_{\omega 1}$ is neither equal to that of a tensile σ_{II} nor that of a tensile σ_{III}, fracture will sometimes occur on a plane which is not the action plane of any of the 3 stresses σ_{II}, σ_{III}, $\tau_{\varpi 1}$. On the actual fracture plane there will then be a ($\sigma_{\perp}^{t}$, $\tau_{\perp\perp}$, $\tau_{\perp\parallel}$) stress combination, called a Mode A*-fracture. This is a major difference to the fracture behavior of brittle isotropic material, where a comparable stressing (a combination of a tensile stress σ^{t} and a shear stress τ) on the fracture plane is not possible.

It seems necessary to have a closer look on the stress distribution. In general, a (σ_{II}, σ_{III}, $\tau_{\varpi 1}$) stress state leads to a stressing $\sigma_n(\theta)$, $\tau_{nt}(\theta)$, $\tau_{n1}(\theta)$ on all planes (parallel to the fiber direction). The maximum of τ_{n1} with the magnitude of $\tau_{n1} = \tau_{\varpi 1}$ occurs on a plane inclined by the angle δ to the action plane of σ_{II}, see Fig. 36. On planes inclined by ±90° to the plane of $\tau_{n1} = \tau_{\varpi 1}$ the longitudinal shear τ_{n1} is zero. In between these extreme inclinations of the fiber parallel action plane τ_{n1} follows a cosine function, compare (Eq. 3).

A rotation of the fiber parallel action plane by ±90° (in the real material) corresponds to a rotation of ±180° in the Mohr's circle. This leads to an illustration of the state of stress as a Mohr's circle with a "cosine-half-wave" on top of the Mohr's circle. The amplitude of the cosine-function is $\tau_{n1} = \tau_{\omega 1}$ (compare Fig. 36). The maximum of τ_{n1} is – with respect to the σ_{II}-axis – rotated by the angle δ in reality and 2δ in the Mohr's circle respectively.

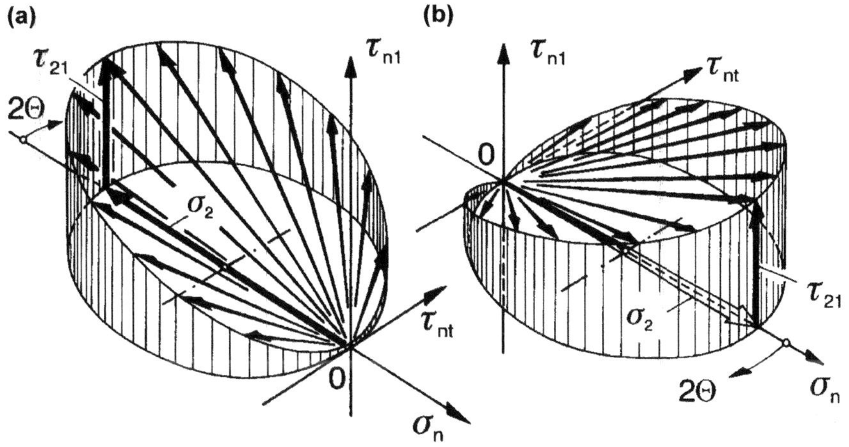

Fig. 37. Spatial vector fan for combined (σ_2, τ_{21})-stressings: a) for $\sigma_2 < 0$; b) for $\sigma_2 > 0$

In addition to (Eq. 32) and (Eq. 33) there is now an additional equation for the longitudinal shear τ_{n1} in dependence of the angle Θ:

$$\tau_{n1}(\Theta) = \tau_{\omega 1}\cos(\Theta - \delta) = \sqrt{\tau_{21}^2 + \tau_{31}^2}\cos(\Theta - \delta) \qquad \text{(Eq. 36)}$$

A stress vector ($\sigma_n(\Theta)$, $\tau_{nt}(\Theta)$, $\tau_{n1}(\Theta)$) in Fig. 36 reaches from the center of origin to a point P on the cosine-half-wave. The infinite number of tips of all stress vectors (acting on planes $-90° \leq (\Theta\text{-}\delta) \leq +90°$) of a given ($\sigma_2$, σ_3, τ_{23}, τ_{31}, τ_{21}) state of stress form together the cosine-half-wave in Fig. 36. All these vectors form a spatial "vector fan", special examples of which are shown in Fig. 37.

The fracture limit for all stress vectors is now no longer an enveloping fracture limit curve in the (σ_n, τ_{nt})-plane, but a surface in the (σ_n, τ_{nt}, τ_{n1})-space. This fracture envelope comprises all cosine-half-waves and corresponding stress vectors which do not lead to fracture. In order to differentiate this fracture envelope from other well known envelopes for instance in the (σ_1, σ_2, τ_{21})-space, the envelope in the (σ_n, τ_{nt}, τ_{n1})-space is called Master Fracture Body (MFB) (compare Fig. 38).

The visualization of the stress state on any fiber parallel action plane by the cosine-half-waves and the visualization of the (σ_n, τ_{nt}, τ_{n1})-fracture limit by the Master Fracture Body (MFB) are very helpful both for the interpretation of the results of a fracture analysis of special stress states and quite generally for a deeper understanding of the background and the results of Puck's criteria.

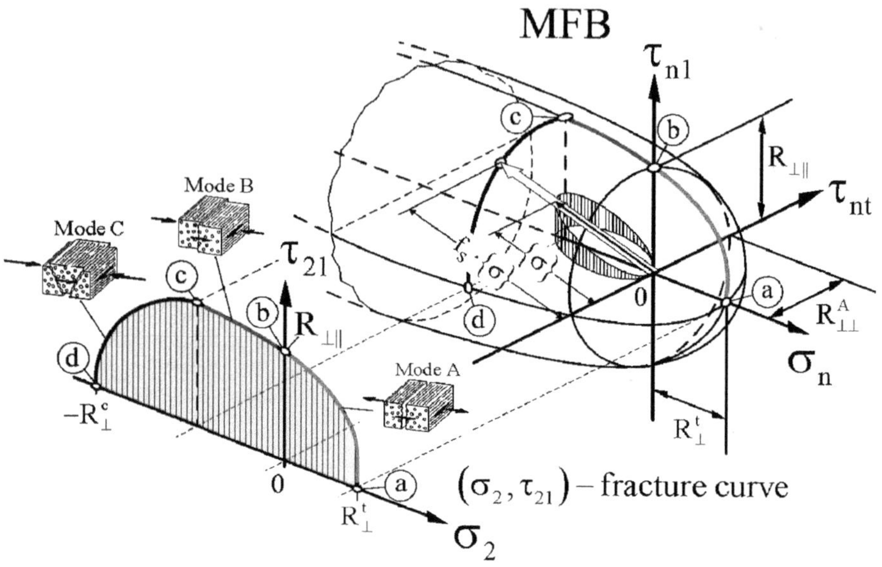

Fig. 38. IFF-Master Fracture Body [Lutz 2006]

It helps for instance to understand that on the surface of the MFB at certain places "dead areas" appear [Ritt 1999]. One of these dead areas contains the "dead branch" between the points S and T in Fig. 31. Stress points (σ_n, τ_{nt}, τ_{n1}) in "dead areas" symbolize (σ_n, τ_{nt}, τ_{n1})-stresses *on the fracture plane* which never occur (The cosine half wave is unable to get in contact with these parts of the surface of the MFB).

All steps of the application of Puck's action plane fracture criteria can be visualized from Fig. 38. In order to answer for instance the question "what stress combinations are sustainable?" for a given (σ_2, τ_{21})-stress combination, one steps into the (σ_n, τ_{nt}, τ_{n1})-stress space where the Master Fracture Body (MFB) is given. Here, one has to find the point on the MFB-surface where the cosine shaped stress line touches the MFB.

What the algorithm of the fracture analysis really performs is the following procedure: Every (σ_n, τ_{nt}, τ_{n1}) stress vector on any inclined section is elongated (stretched) in its original direction by the so called stretch factor $f_S = (f_E)^{-1}$ the way that the tip of the vector just gets in contact with the MFB-surface. The (σ_n, τ_{nt}, τ_{n1})-vector which needs the minimal stretch factor f_{Smin} is the one which in reality leads to IFF. For (σ_2, τ_{21})-combinations Puck has found an analytical solution. For the general (σ_1, σ_2, σ_3, τ_{23}, τ_{31}, τ_{21})-stress combinations fast numerical search procedures have been developed. In Fig. 38 the cosine line in the tensile region of σ_n has a contact point on the line "$\tau_{nt} = 0$", which means $\theta_{fp} = 0°$. In the compressive region the contact point can be in an area where all 3 stresses σ_n, τ_{nt}, τ_{n1} exist. This means that $\theta_{fp} \neq 0°$ is possible. It is interesting to see in Fig. 38 how the fracture curve on the (σ_2, τ_{21})-plane is represented in the MFB. Surprisingly the stress σ_n on the fracture surface is constant ($\sigma_n = -R_{\perp\perp}^A$) in the whole area of oblique fractures ($\theta_{fp} \neq 0°$), see part (c) to (d) of the corresponding fracture curves.

4.2.6 Universal 3-D-formulation of the action plane related IFF-criteria

4.2.6.1 Preliminary remarks

The action plane related IFF-criteria are valid for arbitrary combinations of the stresses σ_1, σ_2, σ_3, τ_{23}, τ_{31}, τ_{21}. The influence stresses parallel to the fibers (σ_1) have on IFF is assumed to be negligible as long as σ_1 is much smaller than the fiber parallel strength $R_{\parallel}$. Thus, that influence is neglected in a first approach and will be discussed later (compare chapter "Influence of stresses σ_1 acting parallel to the fibers on IFF").

Puck's approach is based on the hypothesis that the stresses acting on a fracture surface provoke fracture, this was first proposed by Hashin [Hashin1980]. Consequently, Puck's criteria are formulated with the stresses $\sigma_n(\theta)$, $\tau_{nt}(\theta)$, $\tau_{n1}(\theta)$ (compare Fig. 18) which can easily be calculated from the stresses σ_2, σ_3, τ_{23}, τ_{31}, τ_{21} (see (Eq. 1), (Eq. 2), (Eq. 3)).

The stresses $\sigma_n(\theta)$, $\tau_{nt}(\theta)$, $\tau_{n1}(\theta)$ depend on the angle θ of the plane under consideration or in other words of their action plane. The same is true for the risk of fracture. On a plane with $\theta = \theta_1$ the risk of fracture might be higher than on a plane $\theta = \theta_2$. Eventually, when the lamina stresses are increased to the fracture limit the lamina will suffer IFF on a fracture plane $\theta = \theta_{fp}$. On this plane the risk of fracture is highest and here the stresses $\sigma_n(\theta)$, $\tau_{nt}(\theta)$, $\tau_{n1}(\theta)$ reach first the fracture limit.

A measure for the likelihood of fracture is the stress exposure $f_E(\theta)$ on the action plane. In fact, the plane with the highest stress exposure will be the fracture plane ($f_E(\theta) = [f_E(\theta)]_{max} = f_E\,|_{\theta=\theta fp}$). If the stresses of the UD-lamina (σ_2, σ_3, τ_{23}, τ_{31}, τ_{21}) are multiplied by the smallest stretch factor $f_{Smin} = 1/[f_E(\theta)]_{max}$ the lamina will suffer IFF exclusively on that fracture plane. All other planes remain intact.

For the general 3D-state of stress, the fracture plane has to be found numerically. That means that the stresses $\sigma_n(\theta)$, $\tau_{nt}(\theta)$, $\tau_{n1}(\theta)$ need to be

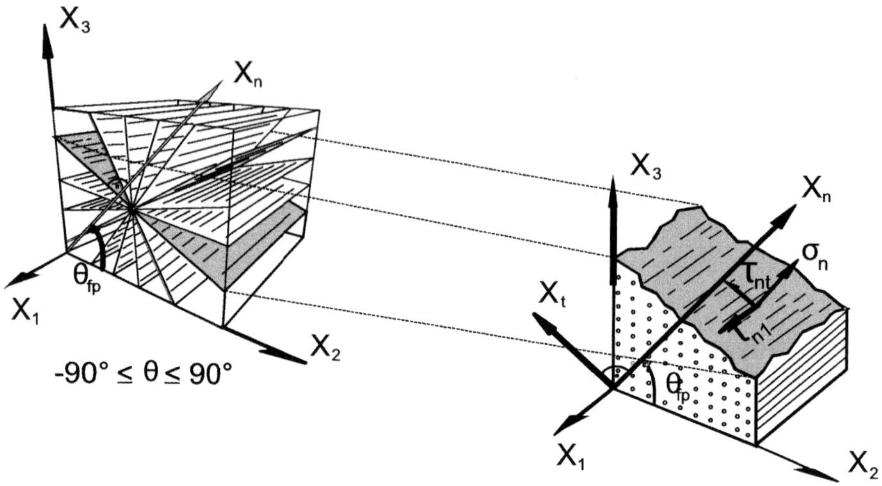

Fig. 39. Search for the fracture plane

calculated for all planes with angles $-90° \leq \theta \leq 90°$ and that for all these angles the stress exposure $f_E(\theta)$ needs to be calculated, see Fig. 39. This is done by inserting the stresses into an IFF-criterion formulated with the stresses of the action plane $\sigma_n(\theta)$, $\tau_{nt}(\theta)$, $\tau_{n1}(\theta)$. The procedure is usually done in 1°-steps leading to 180 calculations. With modern computers that is no problem. Nevertheless, it makes sense to reduce the numerical effort as far as possible.

Figure 40 illustrates the course of the angle dependent stress exposure $f_E(\theta)$ as a function of the angle θ of the action plane for a selected state of stress ($\sigma_1 = \sigma_2 = \sigma_3 = \tau_{21} = 0$; $\tau_{23} = \tau_{31}$). As a result of the fracture analysis one gets a fracture angle $\theta_{fp} = 58°$ and a stress exposure $f_E = 0.5$. This corresponds to the global maximum of the bold curve in Fig. 40. If the stresses are multiplied by the stretch factor $f_S = 1/f_E = 2$, for $\theta = 58°$ $f_E = 1$ is reached in this case. The thin curve in Fig. 40 illustrates the curve $f_E(\theta)$ resulting from multiplying the stresses by $(f_S(\theta_{fp}) = 1/f_E(\theta_{fp}))$.

Originally in 1993, Puck proposed the following IFF-condition for compressive stress σ_n^c:

$$\left(\frac{\tau_{nt}(\theta_{fp})}{R_{\perp\perp}^{A} - p_{\perp\perp}^{c} \cdot \sigma_n(\theta_{fp})}\right)^2 + \left(\frac{\tau_{nl}(\theta_{fp})}{R_{\perp\parallel}^{A} - p_{\perp\parallel}^{c} \cdot \sigma_n(\theta_{fp})}\right)^2 = 1,\ for\ \sigma_n < 0 \qquad \text{(Eq. 37)}$$

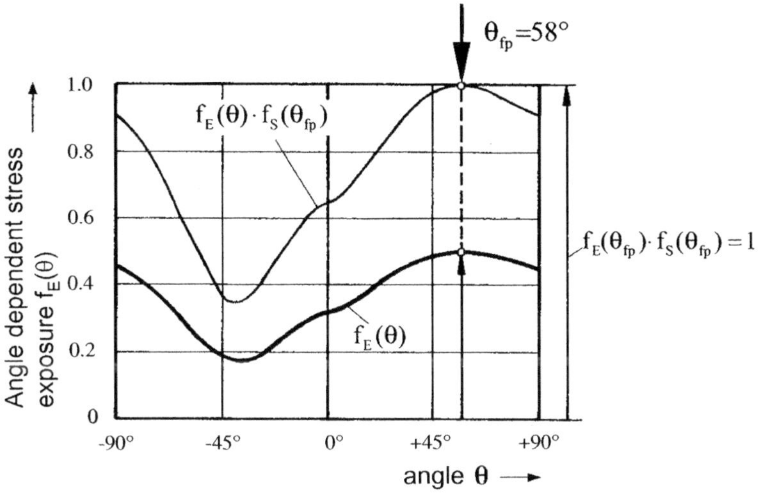

Fig. 40. Stress exposure $f_E(\theta)$ for a τ_{23}/τ_{31}-combination

With the inclination of the (σ_n, τ_{nt})- and (σ_n, τ_{n1})-curves at the point $\sigma_n = 0$:

$$-p_{\perp\perp}^{c} = \left[\frac{\partial \tau_{nt}}{\partial \sigma_n}\right]_{\sigma_n=0} \qquad \text{(Eq. 38)}$$

$$-p_{\perp\|}^{c} = \left[\frac{\partial \tau_{n1}}{\partial \sigma_n}\right]_{\sigma_n=0} \qquad \text{(Eq. 39)}$$

Here, the classical quadratic additive approach was chosen for the interaction of the two shear stresses τ_{nt} and τ_{n1}. The two denominators fulfill Puck's second hypothesis stating that a compressive stress σ_n^c (a stress with a negative value) adds an additional fracture resistance $-p_{\perp\perp}^c\sigma_n(\theta_{fp})$ to $R_{\perp\perp}^A$ and $-p_{\perp\|}^c\sigma_n(\theta_{fp})$ to $R_{\perp\|}^A$, respectively.

At this point one should recall the correct meaning of this statement: It does not mean that adding an additional compressive stress σ_2^c reduces the IFF-stress exposure $f_E = [f_E(\theta)]_{max}$[21]! It does only mean that the stress exposure on the action plane $f_E(\theta)$ under consideration decreases, if $|\sigma_n^c|$ on this plane grows. This implies that on other planes – with a different angle

[21] Even though that might be true in some cases.

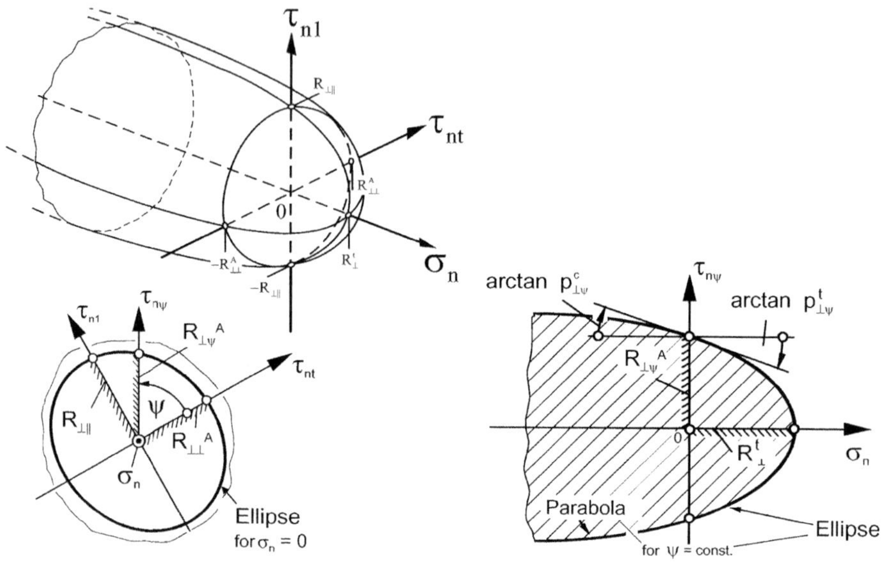

Fig. 41. Master Fracture Body in the (σ_n, τ_{nt}, τ_{n1})-space

θ – where the same state of stress leads to different stresses $\sigma_n(\theta)$, $\tau_{nt}(\theta)$, $\tau_{n1}(\theta)$ the stress exposure might increase.

Another important aspect is the difference between a conventional fracture body written in σ_2, σ_3, τ_{23}, τ_{31}, τ_{21} (and σ_1, if fiber fracture is covered, too) on the one hand and the Master Fracture Body (MFB) describing which stress-combinations (σ_n, τ_{nt}, τ_{n1}) lead to fracture on their action plane.

Again, the difference can well be illustrated regarding the simple example of plane stress ($\sigma_3 = \tau_{23} = 0$). Figure 24 shows the fracture curve for the (σ_2, τ_{21})-stress combination. The curve is closed indicating that any stress ratio σ_2/τ_{21} leads to fracture, if the magnitude of the stresses is sufficiently high. Figure 41 shows schematically the corresponding Master Fracture Body (MFB) in the (σ_n, τ_{nt}, τ_{n1})-space. Additionally, a cross section as described by (Eq. 37) and an arbitrary longitudinal section of the (MFB) are shown.

Comparing the fracture curve in the (σ_2, τ_{21})-diagram and the longitudinal section of the MFB ((σ_n, $\tau_{n\psi}$)-diagram) several misunderstandings and misinterpretations are possible as long as the subject has not been fully understood. First thing to recognize is that any longitudinal section of the MFB has no intersection with the σ_n-axis in the area of $\sigma_n < 0$. At first glance that seems to be in contrast both to common sense and the (σ_2, τ_{21})-fracture curve: Does transverse compression not lead to fracture?

Of course, it does! But not on the action plane of the transverse compressive stress σ_2! A single transverse compressive stress does never ever provoke fracture on its action plane. That is, why the MFB is open to the negative σ_n-axis. For pure transverse compression σ_2^c fracture does occur on a plane inclined by ±54°. That means, that first the stresses $\sigma_n(\theta = \pm 54°)$ and $\tau_{nt}(\theta = \pm 54°)$ have to be calculated. In the case of pure (σ_2, τ_{21})-loading (Eq. 1) simplifies to

$$\sigma_n = \sigma_2 \cdot \cos^2\theta,\ for\ \sigma_3 = \tau_{23} = 0 \qquad \text{(Eq. 40)}$$

$$\tau_{nt} = -\sigma_2 \cdot \sin\theta\cos\theta,\ for\ \sigma_3 = \tau_{23} = 0 \qquad \text{(Eq. 41)}$$

$$\tau_{n1} = \tau_{21} \cdot \cos\theta,\ for\ \tau_{31} = 0 \qquad \text{(Eq. 42)}$$

For pure transverse compression the longitudinal shear stress τ_{21} and consequently τ_{n1} are zero. Thus on the fracture plane $\theta_{fp} = \pm 54°$ just σ_n and τ_{nt} are acting. The magnitude of these stresses can be calculated with (Eq. 40) to (Eq. 42) and is also illustrated in Fig. 25.

This simple example might help to understand the procedure the concept of action plane related fracture criteria implies. The criteria are formulated with the stresses of the action plane. For all possible action planes with inclination angles θ between −90° and +90° the stress exposure $f_E(\theta)$ is calculated and the plane with the highest stress exposure determined. This plane is the fracture plane (compare Fig. 39 and Fig. 40).

The fracture condition (Eq. 37) mirrors well the mechanics and the material behavior. However, it is not very well suited for the numerical search of the fracture plane as explained below.

If a state of stress $\sigma_n(\theta)$, $\tau_{nt}(\theta)$, $\tau_{n1}(\theta)$ does not lead to fracture on the plane examined, the expression on the left hand side of (Eq. 37) is smaller than "1". Fracture will occur on this plane if all three stresses are multiplied by the reciprocal value of the stress exposure $[1/f_E(\theta)]$, the so called stretch factor f_S. This magnification factor leading to the value "1" on the left hand side of (Eq. 37) is unknown at first and must be calculated. However, if f_S is introduced as a factor for σ_n, τ_{nt}, τ_{n1} in (Eq. 37), a 4th grade equation for f_E results. This equation cannot easily be solved. Thus, (Eq. 37) seems to be unsuitable for the calculation of $f_E(\theta)$.

Geometrically (Eq. 37) describes a fracture body with elliptical cross sections for σ_n^c = const. (compare Fig. 41). Of course the fracture body can also mathematically be described by its longitudinal sections, the so called contour lines. By this means the numerical effort for the search of the fracture plane can considerably be reduced. The simple reason for that is the following:

The direction of the stress vector does not change when it is stretched by the stretch factor f_S. This means that the ratios between the three stresses $\sigma_n(\theta):\tau_{nt}(\theta):\tau_{n1}(\theta)$ remain constant. Geometrically, the extension of the stress vector takes place within a longitudinal section characterized by the ratio $\tau_{n1}(\theta)/\tau_{nt}(\theta) = \tan\psi$ (Fig. 41). This means, that $f_E(\theta)$ can relative easily be calculated, if the fracture body is mathematically defined by its longitudinal sections.

In the following the mathematical derivation of the fracture condition is given. Here, of course, a distinction of the cases transverse tension ($\sigma_n > 0$) and transverse compression ($\sigma_n < 0$) is necessary.

4.2.6.2 Fracture condition for tensile stress σ_n

The longitudinal section of the MFB characterized by $\tau_{nt} = 0$ ($\psi = 90°$) is well known from experimental experience. This special section is the same as the (σ_2, τ_{21})-fracture curve. Fracture occurs on the common action plane of σ_2 and τ_{21}, thus $\sigma_n(\theta_{fp}) = \sigma_2$ and $\tau_{n1}(\theta_{fp}) = \tau_{21}$. In the region of tensile stress σ_n on the fracture plane this fracture curve can well be fitted with an elliptical function having a slightly negative inclination at $\sigma_n = 0$, $\tau_{n1} = R_{\perp\parallel}$ and hitting the σ_n-axis perpendicularly at $\sigma_n = R_\perp^t$, $\tau_{n1} = 0$.

Puck assumes that all other longitudinal sections (ψ = const.) can be described with similar curves. The resultant shear stress for such a section is derived from the two stresses $\tau_{nt}(\theta)$ and $\tau_{n1}(\theta)$ with common action plane (compare Fig. 18):

$$\tau_{n\psi}(\theta) = \sqrt{\tau_{nt}^2(\theta) + \tau_{n1}^2(\theta)} \qquad \text{(Eq. 43)}$$

Now, the problem can be treated as 2-dimensional with the two stresses $\sigma_n(\theta)$ (in the following abbreviated with σ_n) and $\tau_{n\psi}(\theta)$ (abbreviated with $\tau_{n\psi}$) similar to that of the classical fracture envelope of Otto Mohr, compare Fig. 31. Defining the fracture resistance of the action plane against a resultant shear stress $\tau_{n\psi}$ as $R^A_{\perp\psi}$ (see Fig. 41), the fracture condition formulated as an elliptic equation results to be:

$$\left(\frac{\tau_{n\psi}}{R^A_{\perp\psi}}\right)^2 + c_1 \cdot \frac{\sigma_n}{R_\perp^{At}} + c_2 \cdot \frac{\sigma_n^2}{\left(R_\perp^{At}\right)^2} = 1 \;,\; for\ \sigma_n \geq 0 \qquad \text{(Eq. 44)}$$

The constants c_1 and c_2 can be determined by boundary conditions: For $\tau_{n\psi} = 0$ the expression $\sigma_n = R_\perp^{At}$ must result. This leads to

$$c_1 + c_2 = 1 \qquad \text{(Eq. 45)}$$

At the intersection with the $\tau_{n\psi}$-axis the fracture curve takes the value $R_{\perp\psi}^A$ and the inclination shall be by definition

$$\left(\frac{\partial \tau_{n\psi}}{\partial \sigma_n} \right)_{\sigma_n = 0} = - p_{\perp\psi}^t \qquad \text{(Eq. 46)}$$

Implicit differentiation of (Eq. 44) at the point $\sigma_n = 0$, $\tau_{n\psi} = R^A_{\perp\psi}$ gives:

$$\frac{2}{R_{\perp\psi}^A} \cdot \left(\frac{\partial \tau_{n\psi}}{\partial \sigma_n} \right)_{\sigma_n = 0} + \frac{c_1}{R_\perp^{At}} = 0 \qquad \text{(Eq. 47)}$$

That leads eventually to the fracture condition:

$$\left(\frac{\tau_{n\psi}}{R_{\perp\psi}^A} \right)^2 + 2 \cdot \frac{p_{\perp\psi}^t \cdot \sigma_n}{R_{\perp\psi}^A} + \left(1 - 2 \cdot \frac{p_{\perp\psi}^t \cdot R_\perp^{At}}{R_{\perp\psi}^A} \right) \cdot \frac{\sigma_n^2}{\left(R_\perp^{At} \right)^2} = 1, \qquad \text{(Eq. 48)}$$

$$for\ \sigma_n \geq 0$$

4.2.6.3 Fracture condition for compressive stress σ_n

In principle several mathematical functions might be adequate to model the Master Fracture Body in the zone of negative σ_n.

Up to a stress magnitude $|\sigma_2^c| \approx 0{,}4R_\perp^c$ (This is the region of $\theta_{fp} = 0$) the experimentally determined (σ_2, τ_{21})-fracture curve can well be modeled by a parabola. The mentioned experimental experience and the fact that Otto Mohr himself has assumed that the increase in sustainable shear stress caused by a superimposed compressive stress σ_n would grow less than linearly with σ_n led Puck to a general parabolic approach for $\sigma_n < 0$. A suitable parabola can be formulated as follows:

$$\left(\frac{\tau_{n\psi}}{R_{\perp\psi}^A} \right)^2 + c \cdot \sigma_n = 1 \ , \ for\ \sigma_n < 0 \qquad \text{(Eq. 49)}$$

Here, the inclination at $\sigma_n = 0$ shall be by definition:

$$\left(\frac{\partial \tau_{n\psi}}{\partial \sigma_n}\right)_{\sigma_n=0} = -\,p^c_{\perp\psi} \qquad \text{(Eq. 50)}$$

Differentiation of (Eq. 49) at $\sigma_n = 0$, $\tau_{n\psi} = 0$ gives with (Eq. 50):

$$c = \frac{2 \cdot p^c_{\perp\psi}}{R^A_{\perp\psi}} \qquad \text{(Eq. 51)}$$

All this leads finally to the following fracture condition for compressive stress σ_n on the fracture plane:

$$\left(\frac{\tau_{n\psi}}{R^A_{\perp\psi}}\right)^2 + 2 \cdot \frac{p^c_{\perp\psi}}{R^A_{\perp\psi}} \cdot \sigma_n = 1 \;\; , for\ \sigma_n < 0 \qquad \text{(Eq. 52)}$$

In (Eq. 48) and (Eq. 52) there are still the fracture resistances $R^A_{\perp\psi}$ and the inclination parameters $p_{\perp\psi}$ as unknown parameters. Consequently, he next paragraphs deal with the definition of reasonable values for these parameters.

4.2.6.4 Description of the cross section of the MFB at $\sigma_n = 0$, determination of $R^A_{\perp\psi}$

Figure 41 is helpful for the interpretation of the parameter $R^A_{\perp\psi}$. In fact, the fracture resistance $R^A_{\perp\psi}$ is – at the point $\sigma_n = 0$ – the distance from the point of origin to the surface of the MFB under the angle ψ. This angle ψ characterizes the longitudinal section and depends on the ratio of τ_{nt} and τ_{n1}. For $\psi = 0°$ ($\tau_{n\psi} = \tau_{n0°} = \tau_{nt}$, $\tau_{n1} = 0$) the fracture resistance is $R^A_{\perp\perp}$ and for $\psi = 90°$ ($\tau_{n\psi} = \tau_{n90°} = \tau_{n1}$, $\tau_{nt} = 0$) it is $R^A_{\perp\parallel}$.

Considerations based on micro-mechanics suggest that these two fracture resistances ($R^A_{\perp\perp}$ and $R^A_{\perp\parallel}$) are not equal but of very similar magnitude. Thus, the classical elliptic approach for intermediate values $R^A_{\perp\psi}$ seems to be appropriate:

$$\left(\frac{\tau_{n\psi 0}}{R^A_{\perp\psi}}\right)^2 = \left(\frac{\tau_{nt0}}{R^A_{\perp\perp}}\right)^2 + \left(\frac{\tau_{n10}}{R^A_{\perp\parallel}}\right)^2 = 1 \qquad \text{(Eq. 53)}$$

Here, the additional index “0” indicates that all stresses are taken at $\sigma_n = 0$. With the correlations $\tau_{nt0} = \tau_{n\psi 0}\cos\psi$ and $\tau_{n10} = \tau_{n\psi 0}\sin\psi$ (Eq. 53) leads to:

$$\left(\frac{1}{R^A_{\perp\psi}}\right)^2 = \left(\frac{\cos\psi}{R^A_{\perp\perp}}\right)^2 + \left(\frac{\sin\psi}{R^A_{\perp\parallel}}\right)^2 = 1 \qquad \text{(Eq. 54)}$$

However, the normal stress σ_n has no influence on ψ. For all values of σ_n the following correlations are valid:

$$\cos\psi = \frac{\tau_{nt}}{\tau_{n\psi}} \qquad \text{(Eq. 55)}$$

and

$$\sin\psi = \frac{\tau_{n1}}{\tau_{n\psi}} \qquad \text{(Eq. 56)}$$

Thus, (Eq. 53) is generally valid for all σ_n:

$$\left(\frac{\tau_{n\psi}}{R^A_{\perp\psi}}\right)^2 = \left(\frac{\tau_{nt}}{R^A_{\perp\perp}}\right)^2 + \left(\frac{\tau_{n1}}{R^A_{\perp\parallel}}\right)^2 \qquad \text{(Eq. 57)}$$

With (Eq. 57) the Master Fracture Body is defined by a transverse section ($\sigma_n = 0$) and an infinite number of longitudinal sections $\psi = \text{const.}$ ((Eq. 48) and (Eq. 52)). The only parameter left to be defined is the inclination of the longitudinal sections at $\sigma_n = 0$ (compare Fig. 41). Guidelines concerning the choice of values for the inclination parameters can be found in the next paragraphs.

In the case $\tau_{nt} = \tau_{n1} = 0$ it does not matter which value is taken for $p_{\perp\psi}^{t}/R^A_{\perp\psi}$ since this coefficient can be (in the special case of no shear stress) eliminated from (Eq. 48). The fracture condition of the action plane for negative σ_n (Eq. 52) is not defined for $\tau_{nt} = \tau_{n1} = 0$, because there is no fracture on a plane with pure transverse compression. In this case fracture takes place on another action plane.

4.2.6.5 Rearrangement of the fracture condition for searching the fracture plane

To use the action plane related fracture condition

$$F\big(\sigma_n(\theta_{fp}), \tau_{nt}(\theta_{fp}), \tau_{n1}(\theta_{fp})\big) = 1 \qquad \text{(Eq. 58)}$$

first the fracture angle θ_{fp} needs to be determined. The fracture plane (and with this the fracture angle θ_{fp}) is characterized as the action plane with the maximum "local" stress exposure: $f_E(\theta_{fp}) = [f_E(\theta)]_{max}$ with $-90° \le \theta \le 90°$. It is better to use the stress exposure and not the reciprocal stretch factor $f_S = 1/f_E(\theta)$ for the numerical search of the fracture plane, because on planes being free of stress, $[1/f_E(\theta)]$ equals "∞" whereas $f_E(\theta)$ equals "0" which poses no numerical problems.

Assuming that the stresses $\tau_{n\psi}(\theta)$ and $\sigma_n(\theta)$ in (Eq. 48) and (Eq. 52) do not yet cause fracture, they must be multiplied by f_S or divided by $f_E(\theta)$ in order to lead to fracture (on this very plane) and to fulfill the fracture condition. Keeping this in mind, it is obvious that the fracture conditions (Eq. 48) and (Eq. 52) lead to quadratic equations for $f_E(\theta)$, because the stresses occur just in the first and second order in the fracture conditions. The solution of these quadratic equations can be explicitly determined, see (Eq. 12). The resulting expression for $f_E(\theta)$ is homogeneous of first grade concerning the stresses. This means that the stress exposure $f_E(\theta)$ grows linearly with the stresses. Thus the stress exposure is a direct measure of the risk of fracture.

In Fig. 42 all equations needed for the described numerical search of the fracture plane are summarized.

$$f_E(\theta) = \sqrt{\left[\left(\frac{1}{R_\perp^t} - \frac{p_{\perp\psi}^t}{R_{\perp\psi}^A}\right)\cdot\sigma_n(\theta)\right]^2 + \left(\frac{\tau_{nt}(\theta)}{R_{\perp\perp}^A}\right)^2 + \left(\frac{\tau_{n1}(\theta)}{R_{\perp\parallel}}\right)^2} + \frac{p_{\perp\psi}^t}{R_{\perp\psi}^A}\sigma_n(\theta)$$

for $\sigma_n \ge 0$

$$f_E(\theta) = \sqrt{\left(\frac{\tau_{nt}(\theta)}{R_{\perp\perp}^A}\right)^2 + \left(\frac{\tau_{n1}(\theta)}{R_{\perp\parallel}}\right)^2 + \left(\frac{p_{\perp\psi}^c}{R_{\perp\psi}^A}\sigma_n(\theta)\right)^2} + \frac{p_{\perp\psi}^c}{R_{\perp\psi}^A}\sigma_n(\theta)$$

for $\sigma_n < 0$

$$\frac{p_{\perp\psi}^{t,c}}{R_{\perp\psi}^A} = \frac{p_{\perp\perp}^{t,c}}{R_{\perp\perp}^A}\cdot\cos^2\psi + \frac{p_{\perp\parallel}^{t,c}}{R_{\perp\parallel}^A}\cdot\sin^2\psi \qquad \cos^2\psi = \frac{\tau_{nt}^2}{\tau_{nt}^2 + \tau_{n1}^2}$$

$$R_{\perp\perp}^A = \frac{R_\perp^c}{2\cdot\left(1 + p_{\perp\perp}^c\right)} \qquad \sin^2\psi = \frac{\tau_{n1}^2}{\tau_{nt}^2 + \tau_{n1}^2}$$

Fig. 42. Equations needed for the numerical search of the fracture plane

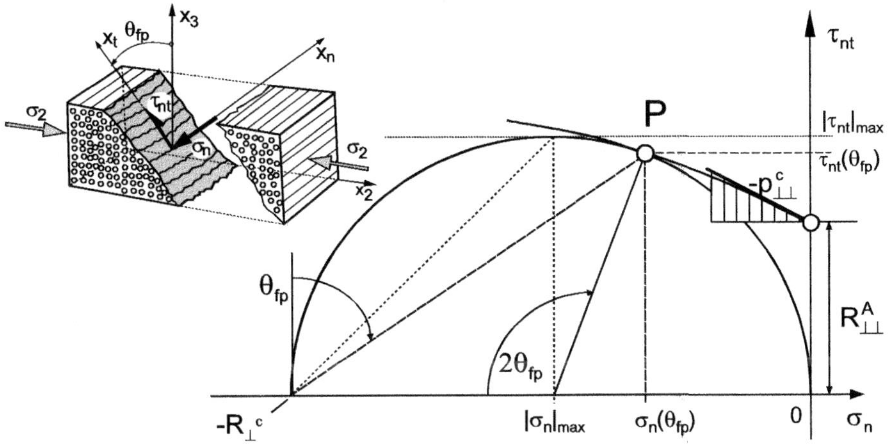

Fig. 43. Derivation of the relation between $R_{\perp\perp}{}^{A}$ and $R_{\perp}{}^{c}$

The equation for $R_{\perp\perp}{}^{A}$ in Fig. 42 can be derived geometrically from Fig. 43. The coordinates of the point P in Fig. 43 are

$$\sigma_n = -\frac{R_{\perp}^{c}}{2}\left[1-\sin\left(2\theta_{fp}^{c}-90°\right)\right] \tag{Eq. 59}$$

$$\tau_{nt} = \frac{R_{\perp}^{c}}{2}\left[\cos\left(2\theta_{fp}^{c}-90°\right)\right] \tag{Eq. 60}$$

For these coordinates not only the equation of Mohr's circle for uniaxial compression is valid, but also the equation for the parabolic envelope:

$$\tau_{nt}^{2} = \left(R_{\perp\perp}^{A}\right)^{2} - 2p_{\perp\perp}^{c}R_{\perp\perp}^{A}\sigma_n \tag{Eq. 61}$$

In point P circle and parabola have the same slope:

$$-\tan\left(2\theta_{fp}-90°\right) = -\frac{p_{\perp\perp}^{c}R_{\perp\perp}^{A}}{\tau_{nt}} = -p_{\perp\perp}^{c}\sqrt{\frac{R_{\perp\perp}^{A}}{R_{\perp\perp}^{A}-2p_{\perp\perp}^{c}\sigma_n}} \tag{Eq. 62}$$

From (Eq. 61) and (Eq. 62) follows:

$$\cos\left(2\theta_{fp}^{c}\right) = -\frac{p_{\perp\perp}^{c}}{1+p_{\perp\perp}^{c}} \quad \text{and} \tag{Eq. 63}$$

$$R_{\perp\perp}^{A} = \frac{R_{\perp}^{c}}{2\left(1+p_{\perp\perp}^{c}\right)} \tag{Eq. 64}$$

4.2.6.6 Choice of inclination parameters $p^t_{\perp\psi}$ and $p^c_{\perp\psi}$

The parameters $p_{\perp\parallel}^t$ and $p_{\perp\parallel}^c$ ($\psi = 90°$) can be deduced from the (σ_2, τ_{21})-fracture curve (compare Fig. 24). This is possible due to the fact, that the fracture angle θ_{fp} is zero for both $\sigma_2 > 0$ and $\sigma_2 < 0$. Thus, the stresses on the fracture plane are $\sigma_n = \sigma_2$ and $\tau_{n1} = \tau_{21}$.

Those – experimentally determined – values for the parameters $p_{\perp\parallel}^t$ and $p_{\perp\parallel}^c$ usually lie between 0.25 and 0.35 (Table 1). Quite often experimental results are best fit to (Eq. 48) and (Eq. 52) respectively, if $p_{\perp\parallel}^t$ is chosen slightly higher than $p_{\perp\parallel}^c$. Generally speaking, values between 0.2 and 0.35 lead to reasonable fracture curves (Puck et al. 2002). Table 1 shows typical values for unidirectional layers with thermosetting matrix. They represent mean values resulting from two major experimental projects [Cuntze et al. 1997; Kopp2000].

Corresponding validated data for $p_{\perp\perp}^c$ and $p_{\perp\perp}^t$ ($\psi = 0°$) do not exist, because the (τ_{nt}, σ_n)-fracture curve is experimentally non-accessible in the region $\sigma_n \approx 0$ [Puck 1996, Ritt 1999]. However, $p_{\perp\perp}^c$ can indirectly be drawn from transverse compression fracture tests recording the fracture angle θ_{fp}. Here the following equation holds:

$$p^c_{\perp\perp} = \frac{1}{2 \cdot \cos^2 \theta_{fp}} - 1 \qquad \text{(Eq. 65)}$$

As stated before, for both GFRP and CFRP one obtains – under uniaxial transverse compression – fracture angles slightly above 50° from experiments. In this case values greater than $p_{\perp\perp}^c = 0.21$ result from (Eq. 65). However, it should be recognized that the measurement of fracture angles is very difficult and that results scatter considerably [Cuntze et al. 1997; Kopp2000]. Moreover, (Eq. 65) is very sensitive even to marginal changes of the fracture angle θ_{fp} under pure transverse compression. Thus, it cannot be recommended to use (Eq. 65) for fitting $p_{\perp\perp}^c$ and it must be accepted for the moment that no experiments are available for the validated determination of $p_{\perp\perp}^c$ and $p_{\perp\perp}^t$.

At first glance it seems logical to choose $p_{\perp\perp}^t$ and $p_{\perp\perp}^c$ identical to $p_{\perp\parallel}^t$ and $p_{\perp\parallel}^c$, respectively, because both $\tau_{\perp\parallel}$ and $\tau_{\perp\perp}$ are shear stressings in a plane parallel to the fibers. In fact, this approach is reasonable. However, from a micro-mechanical point of view $\tau_{\perp\parallel}$- and $\tau_{\perp\perp}$-stressing lead to different stress distributions in the matrix. In the first case the shear stressing is oriented parallel to the fibers, in the latter case it is transverse. This has consequences for the micro-cracking preceding IFF, too. Consequently the values for $R_{\perp\parallel}$ and $R_{\perp\perp}^A$ are slightly different for both cases. Likewise, for $p_{\perp\parallel}^c$ and $p_{\perp\perp}^c$ similar but not necessarily equal values can be expected. The same is valid for $p_{\perp\parallel}^{t}$ and $p_{\perp\perp}^{t}$. However, there is no reason for $p_{\perp\parallel}^{c} = p_{\perp\parallel}^{t}$ since in the corresponding longitudinal cut through the master fracture body σ_n changes its fracture mode from impeding IFF-formation ($\sigma_n < 0$) to promoting IFF-formation ($\sigma_n > 0$).

Evaluating all available information it is hereby recommended to use the set of parameters given in Table 1.

Table 1. Inclination parameters for typical FRP [Cuntze et al. 1997; Kopp 2000; Puck et al. 2002]

$\varphi = 60\ \%$	$p_{\perp\parallel}^{t}$ [–]	$p_{\perp\parallel}^{c}$ [–]	$p_{\perp\perp}^{t}$ [–]	$p_{\perp\perp}^{c}$ [–]
GFRP/ Epoxy	0.3	0.25	0.20 to 0.25	0.20 to 0.25
CFRP/ Epoxy	0.35	0.3	0.25 to 0.30	0.25 to 0.30

In order to keep the MFB at all points continuously differentiable an interpolation is necessary for all longitudinal sections $\psi \neq 0°$ and $\psi \neq 90°$. This is done as follows:

$$\frac{p_{\perp\psi}^{t,c}}{R_{\perp\psi}^{A}} = \frac{p_{\perp\perp}^{t,c}}{R_{\perp\perp}^{A}} \cdot \cos^2\psi + \frac{p_{\perp\parallel}^{t,c}}{R_{\perp\parallel}^{A}} \cdot \sin^2\psi \qquad \text{(Eq. 66)}$$

with

$$\cos^2\psi = \frac{\tau_{nt}^2}{\tau_{nt}^2 + \tau_{n1}^2} \qquad \text{(Eq. 67)}$$

$$\sin^2\psi = 1 - \cos^2\psi \qquad \text{(Eq. 68)}$$

4.2.6.7 Limits of validity of the recommended inclination parameters

There is one last factor of relevance with regard to the choice of the inclination parameters $p_{\perp\perp}^{t}$ and $p_{\perp\perp}^{c}$. This is due to the fracture mechanism of brittle materials subjected to a pure τ_{23}-stress (Fig. 44).

Intrinsically brittle materials subjected to a pure τ_{23}-stress always fail due to the principal normal stress $\sigma_n = \sigma_{II}$ occurring on a fracture plane which is inclined towards the direction of the τ_{23}-stress by $\theta_{fp} = 45°$ [Paul 1961].

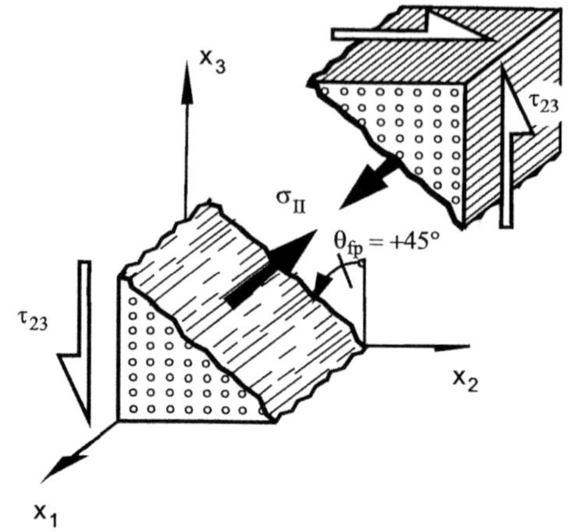

Fig. 44. Fracture due to pure τ_{23}-stress

The fracture criterion will only predict this behavior and not a 'mixed mode fracture' due to simultaneously acting τ_{nt}- and σ_n^t-stresses, if the following condition is complied [Cuntze et al. 1997]:

$$p_{\perp\perp}^t \leq \frac{R_{\perp\perp}^A}{R_\perp^t} - \frac{R_\perp^t}{R_{\perp\perp}^A} \qquad \text{(Eq. 69)}$$

This has also been shown by numerical studies [Ritt 1999]. In case of $p_{\perp\perp}^c = p_{\perp\perp}^t = p_{\perp\perp}$ – as proposed above – and applying the equation in Fig. 42 to replace $R_{\perp\perp}^A$ one gets

$$p_{\perp\perp} \leq \frac{\sqrt{(r+4)^2 + 2\cdot(r^2-4)\cdot(r+2)} - (r+4)}{2\cdot(r+2)} \qquad \text{(Eq. 70)}$$

with $r = R_\perp^c / R_\perp^t$.

In accordance with Table 1 usually $p_{\perp\perp} = 0.25$ is chosen for GFRP and $p_{\perp\perp} = 0.3$ for CFRP. Under terms of (Eq. 70) this does not cause any problems for CFRP as long as $R_\perp^c/R_\perp^t \geq 3$ holds, which is a valid assumption for most CFRP having a thermosetting matrix. As long as $R_\perp^c/R_\perp^t \geq 2.8$ holds, $p_{\perp\perp} = 0.25$ is applicable. For most but not for all GFRP having a thermosetting matrix this ratio will not be undercut. For the remaining exceptional cases it makes sense to determine the value of $p_{\perp\perp}$ according to (Eq. 70). However, in accordance with Table 1 no values lower than $p_{\perp\perp} = 0.2$ (corresponding to $R_\perp^c/R_\perp^t = 2.65$) should be used since otherwise unrealistic transverse compression fracture angles θ_{fp}^c would be calculated from (Eq. 65).

If in case of $R_\perp^c/R_\perp^t < 2.65$ the lower limit $p_{\perp\perp} = 0.2$ is chosen one has to accept that the calculation procedure incorporating (Eq. 48) and (Eq. 52) will not lead to a fracture angle of $\theta_{fp} = 45°$ for a pure τ_{23}-stress. In other words the fracture criterion will indicate a 'mixed mode fracture' due to simultaneously acting shear stress τ_{nt}- and **tensile** σ_n-stress.

However, all FRP known to the author have a ratio of $R_\perp^c/R_\perp^t > 2.65$. This applies according to latest measurements to carbon fiber reinforced PEEK ($R_\perp^c = 270$ N/mm^2; $R_\perp^t = 90$ N/mm^2) and glass fiber reinforced Polyamid12 ($R_\perp^c = 141$ N/mm^2; $R_\perp^t = 42$ N/mm^2), too.

But – on the other hand – it can be expected that for a material with an unusually low ratio $R_\perp^c/R_\perp^t$ a mixed mode fracture and the corresponding fracture angle $\theta_{fp} \neq 45°$ resulting from Puck's fracture criteria will really occur.

4.2.7 Analytical 2-D-formulation for plane states of stress

In thin laminates there are often no considerable stresses present in thickness direction. In this case a 2D stress and strength analysis is sufficient and no numerical search of the fracture plane is required. The equations for the stresses on the action plane σ_n, τ_{nt} and τ_{nl} (compare (Eq. 1), (Eq. 2), (Eq. 3)) simplify to:

$$\sigma_n(\theta) = \sigma_2 \cdot \cos^2\theta \qquad \text{(Eq. 40)}$$

$$\tau_{nt}(\theta) = -\sigma_2 \cdot \sin\theta \cdot \cos\theta \qquad \text{(Eq. 41)}$$

$$\tau_{nl}(\theta) = \tau_{21} \cdot \cos\theta \qquad \text{(Eq. 42)}$$

This case has been discussed before and used for the explanation of the different IFF fracture modes (compare chapter "Different IFF-fracture modes" and Fig. 19). Figure 45 illustrates again the (σ_2, τ_{21})-fracture curve and the correspondent lines on the surface of the Master Fracture Body (MFB).

For deriving the analytical 2D-formulation of the IFF-criterion a separate inspection of the three IFF-Modes is useful.

If σ_2 is a tensile stress, the fracture occurs on the plane perpendicular to the x_2-axis ($\theta_{fp} = 0°$) and the fracture mode is Mode A. The stresses on the fracture plane are identical to the stresses σ_2, τ_{21} of the UD-lamina: $\sigma_n = \sigma_2$, $\tau_{nt} = 0$ and $\tau_{nl} = \tau_{21}$. The IFF-criterion for $\sigma_n \geq 0$ (Mode A, compare Fig. 42) can therefore be written with the stresses σ_2 and τ_{21}:

$$f_{E,IFF} = \sqrt{\left[\left(\frac{1}{R_\perp^t} - \frac{p_{\perp\parallel}^t}{R_{\perp\parallel}}\right) \cdot \sigma_2\right]^2 + \left(\frac{\tau_{21}}{R_{\perp\parallel}}\right)^2} + \frac{p_{\perp\parallel}^t}{R_{\perp\parallel}} \sigma_2 \;, \; for\; \sigma_2 \geq 0 \qquad \text{(Eq. 71)}$$

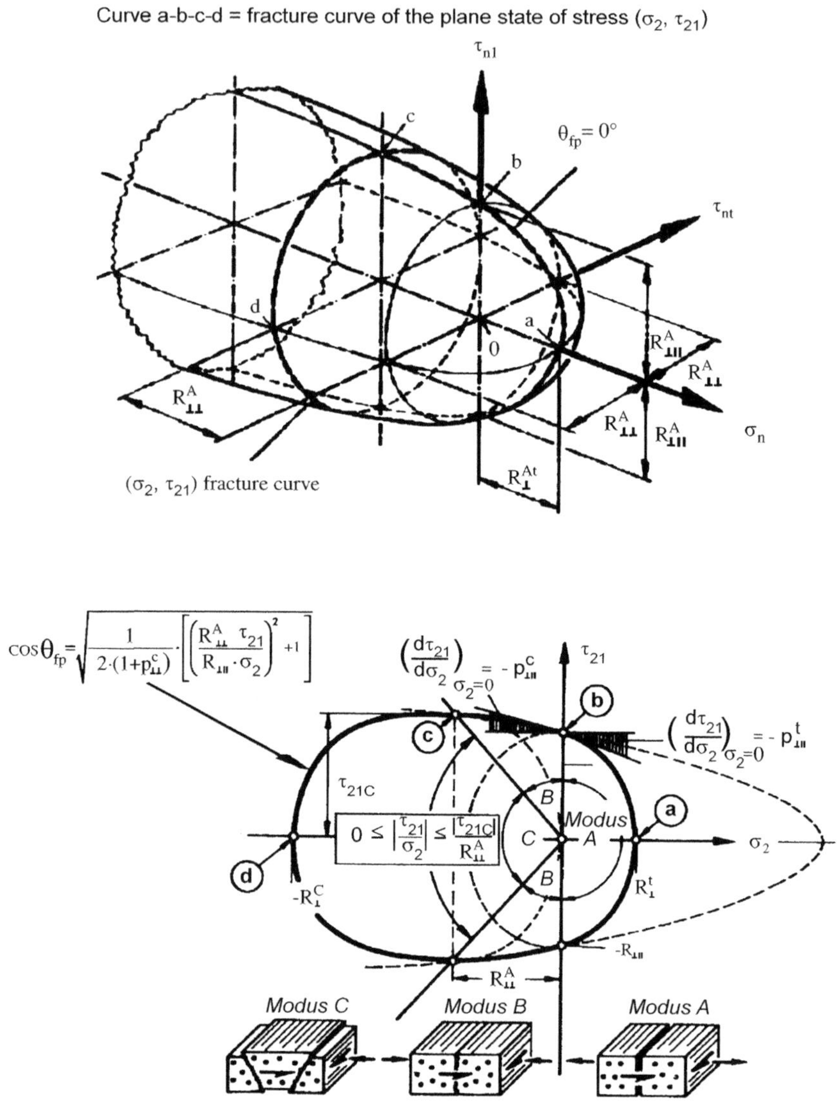

Fig. 45. (σ_2, τ_{21})-fracture curve and its derivation from the Master-Fracture-Body

For $\sigma_n < 0$ the IFF-criterion in Fig. 42 has the general form

$$f_{E,IFF}(\theta) = \sqrt{\left(\frac{\tau_{nt}(\theta_{fp})}{R^A_{\perp\perp}}\right)^2 + \left(\frac{\tau_{n1}(\theta_{fp})}{R_{\perp\parallel}}\right)^2 + \left(\frac{p^c_{\perp\psi}}{R^A_{\perp\psi}}\sigma_n(\theta_{fp})\right)^2} + \frac{p^c_{\perp\psi}}{R^A_{\perp\psi}}\sigma_n(\theta_{fp}), \quad for\ \sigma_n < 0$$

(Eq. 72)

The angle θ_{fp} of the fracture plane is unknown in first place. However, as long as the ratio of the transverse stress at fracture and the transverse compressive strength $|\sigma_2/R_\perp^c|$ does not exceed a critical value of roughly 0.4 the fracture angle equals zero ($\theta_{fp} = 0$) just as for transverse tensile stress. The fracture mode is Mode B and the fracture condition (Eq. 72) can also be written with the stresses σ_2 and τ_{21}. In this case the transverse shear stress is zero ($\tau_{nt} = 0$), the inclination parameter $p_{\perp\psi}^c = p_{\perp\|}^c$ and the shear strength of the action plane $R^A_{\perp\psi} = R_{\perp\|}$ (compare Fig. 41).

$$f_{E,IFF} = \sqrt{\left(\frac{\tau_{21}}{R_{\perp\|}}\right)^2 + \left(\frac{p_{\perp\|}^c}{R_{\perp\|}}\sigma_2\right)^2} + \frac{p_{\perp\|}^c}{R_{\perp\|}}\sigma_2 \ , \sigma_2 < 0 \wedge \left|\frac{\sigma_2}{\tau_{21}}\right| \le \left|\frac{R_{\perp\perp}^A}{\tau_{21,c}}\right| \qquad \text{(Eq. 73)}$$

The fact that the transverse compressive strength $R_\perp^c$ does not occur in (Eq. 73) reflects that $\text{IFF}^{\text{Mode B}}$ is a pure shear fracture on the action plane of τ_{21}. This shear fracture is impeded by the compressive stress σ_2^c acting simultaneously to τ_{21} on the fracture plane. Higher shear stress τ_{21} can be sustained in the presence of a compressive stress σ_2^c. This means that the fracture resistance of the action plane ($\theta = 0°$) is increased by increasing $|\sigma_2^c|$.

If $|\sigma_2^c|$ exceeds a critical value, the fracture resistance of the plane with $\theta = 0°$ is no longer the lowest of all action planes. Consequently, fracture takes place on a different plane with $\theta \neq 0°$. The critical value for $|\sigma_2^c|$ is $R_{\perp\perp}^A$, the fracture resistance of the action plane against transverse shear stressing. The point $|\sigma_2^c| = R_{\perp\perp}^A$ marks the transition point to Mode C. The shear stress reached at this point is $\tau_{21,c}$. Thus the point

$$\sigma_2 = -R_{\perp\perp}^A \ ; \tau_{21,c} = R_{\perp\|} \cdot \sqrt{1 + 2p_{\perp\perp}^c} \qquad \text{(Eq. 74)}$$

marks on the fracture curve the transition from Mode B to Mode C. This will be explained in more detail in the following.

To find an analytical solution in σ_2 and τ_{21} for the fracture condition in Mode C is much more difficult than to formulate the (σ_2, τ_{21})-fracture conditions for Mode A and Mode B. In a first step the extremum problem $d(f_E)/d\theta = 0$ has to be solved. How the problem has been tackled by Puck can be found in [Puck 1996, Puck and Schürmann 1998].

For mathematical simplification Puck coupled the two (usually independent) parameters $p_{\perp\|}^c$ and $p_{\perp\perp}^c$ in the following way:

$$\frac{p_{\perp\perp}^c}{R_{\perp\perp}^A} = \frac{p_{\perp\|}^c}{R_{\perp\|}} \qquad \text{(Eq. 75)}$$

Constraining the free choice of parameters by this means does not considerably change the results of fracture analysis nor does it lead to contradictions or a confinement of the physical basis of Puck's criteria [Puck 1996].

As a provisional result he found a surprising correlation of the fracture angle θ_{fp} and the normal stress $[\sigma_n(\theta_{fp})]_{fr}$ at fracture [Puck 1996, Puck and Schürmann 1998]

$$\left[\sigma_n\left(\theta_{fp}\right)\right]_{fr} = \left(\sigma_2\right)_{fr} \cdot \cos^2\theta_{fp} = -R^A_{\perp\perp} \qquad \text{(Eq. 76)}$$

That means that for the entire range of (σ_2, τ_{21})-combinations which cause fracture by Mode C the stress $[\sigma_n(\theta_{fp})]_{fr}$ on the fracture plane is constant and can be calculated from $[\sigma_n(\theta_{fp})]_{fr} = -R_{\perp\perp}^A = -R_\perp^c / [2(1+ p_{\perp\perp}^c)] \approx -0.4\ R_\perp^c$. That means that all fracture points $(\sigma_2, \tau_{21})_{fr}$ on the Master Fracture Body are located on the boundary of a cross section of the MFB. This is the curve (c) → (d) in Fig. 45.

With these results the analytical solution in (σ_2, τ_{21}) for the fracture criterion of Mode C is

$$f_{E,IFF} = \left[\left(\frac{\tau_{21}}{2\left(1+p^c_{\perp\perp}\right)R_{\perp\parallel}}\right)^2 + \left(\frac{\sigma_2}{R^c_\perp}\right)^2\right]\frac{R^c_\perp}{-\sigma_2},$$

$$\text{for } \sigma_2 < 0 \wedge \left|\frac{\sigma_2}{\tau_{21}}\right| \geq \left|\frac{R^A_{\perp\perp}}{\tau_{21,c}}\right| \qquad \text{(Eq. 77)}$$

Using (Eq. 77) in order to formulate $\cos^2\theta_{fp} = -R_{\perp\perp}^A / (\sigma_2)_{fr}$ the following relationship for the fracture angle θ_{fp} is found

$$\cos\theta_{fp} = \sqrt{\frac{1}{2\cdot\left(1+p^c_{\perp\perp}\right)}\cdot\left[\left(\frac{R^A_{\perp\perp}\cdot\tau_{21}}{R_{\perp\parallel}\cdot\sigma_2}\right)^2+1\right]} \qquad \text{(Eq. 78)}$$

This relation is very useful. For instance it allows to calculate the fracture angle without having calculated the fracture stress $(\sigma_2)_{fr}$ first. It depends only on the ratio of τ_{21}/σ_2. Further it demonstrates that θ_{fp} does not change if $R_{\perp\perp}^A$ and $R_{\perp\parallel}$ are decreased by the same factor for instance by a weakening factor η_{w1} which will be dealt with in the next Section.

4.3 Extensions to the IFF-criteria

The following two chapters are in main parts identical to the corresponding chapters of the annex to VDI 2014 Part 3 [VDI 2006]. These chapters of the VDI 2014 have been written by Alfred Puck and Günther Lutz and the use of this work is authorized by VDI.

4.3.1 Inclusion of stresses not acting on the fracture plane in the action-plane-related inter-fiber fracture criteria

The action-plane-related strength criteria for inter-fiber fracture (IFF) – Fig. 42 and also (Eq. 71), (Eq. 73) and (Eq. 77) – are based on Mohr's hypothesis. According to this, only the stresses $\sigma_n(\theta_{fp})$, $\tau_{nt}(\theta_{fp})$ and $\tau_{n1}(\theta_{fp})$ acting on the parallel-to-fiber *fracture plane* (angle of inclination θ_{fp}) are of decisive importance for fracture stresses at IFF. All parallel-to-fiber section planes are in principle potential fracture planes for IFF. Their angles of inclination θ (definition of θ in Fig. 18) vary between $\theta = -90°$ and $\theta = +90°$, where the section planes with $\theta = -90°$ and $\theta = +90°$ are identical. The fracture plane which is to be expected is determined by finding the section plane with the highest stress exposure $f_E(\theta)_{max}$ dependent on the angle of intersection (The stress exposure is defined in Fig. 21 and Fig. 22). The search for the angle of the fracture plane was analytically anticipated when equations (Eq. 71), (Eq. 73) and (Eq. 77) were set up. If the equations of Fig. 42 are used, the section plane with the highest IFF stress exposure $f_E(\theta_{max}) = f_E(\theta_{fp})$ must be found by a numerical search of the intersection-angle range from $\theta = -90°$ to $\theta = +90°$.

If the fracture plane has been identified in this way, the stresses $(\sigma_2, \sigma_3, \tau_{23}, \tau_{31}, \tau_{21})_{fr}$ when the IFF occurs may be obtained by dividing the components of the given stress $(\sigma_2, \sigma_3, \tau_{23}, \tau_{31}, \tau_{21})$ by the stress exposure $f_E(\theta_{fp})$ present on the fracture plane. This therefore means that it does not matter whether the stresses $\sigma_n(\theta)$, $\tau_{n1}(\theta)$, $\tau_{nt}(\theta)$ acting on other sections with angles $\theta \neq \theta_{fp}$ bring about relatively high or low IFF stress exposures. On account of the gradual development of an IFF due to progressive formation of micro cracks and for probabilistic reasons, it is, however, to be expected that there are stress states where the Mohr approach gives results which in dimensioning calculations would tend to be on the non-conservative side. This problem will be the subject of chapter 4.3.1.2.

According to Mohr's hypothesis, even the parallel-to-fiber stress σ_1 does not have any influence on the IFF since it does not act on a parallel-to-fiber section plane and thus does not act on a potential fracture plane for IFF.

Micromechanical aspects do mean that even in this case a correction will be needed. Chapter 4.3.1.1 will deal with a correction of this kind.

4.3.1.1 Inclusion of a parallel-to-fiber stress σ_1 in the action-plane IFF conditions

Physical considerations

According to Mohr's hypothesis, the stress σ_1 does not have any influence on the IFF due to the fact that the action plane of σ_1 is perpendicular to the plane which is acted on by the stresses σ_n, τ_{nl} and τ_{nt} which are of decisive importance to the IFF. However, a series of effects appear to make it necessary to include a σ_1 term in the IFF criteria which reduces 'IFF strength' somewhat [Puck 1996]. The most essential effect may be the following one. FF is taken to mean the fracture of a very large number of elementary fibers which in turn causes a lamina to loose some of its load-bearing capacity in the fiber direction over a 'macro-region'. Statistical laws, however, state that in the event of tensile stress σ_1 some elementary fibers will already have ruptured before the FF limit of the UD lamina at $|\sigma_1| = R_{\parallel}^t$ has been reached. In the case of compressive stress σ_1 it is possible that individual bundles of fibers could already start kinking before total fracture occurs when $|\sigma_1| = R_{\parallel}^c$. Occasional 'micro fiber fractures' cause local damage in the UD lamina which takes the form of debonding at fiber matrix interfaces and of micro-fractures in the matrix material. They weaken the fiber matrix cohesion and thus also reduce its resistance to IFF.

Analytical treatment

In order to include in IFF analysis the weakening effect resulting from σ_1 in a way which is appropriate to the physical circumstances, the action-plane-fracture resistances $R_{\perp}^t$, $R_{\perp\perp}^A$, $R_{\perp\parallel}$ are multiplied by a degradation factor η_{w1} (w = weakening, 1 $\hat{=}$ weakening due to σ_1). A factor of this kind can be used not only in the tensile range but also in the compressive range of σ_1. For the sake of simplicity it is assumed that the 'weakening factor' η_{w1} has the same numerical value for all three action plane-fracture resistances $R_{\perp}^t$, $R_{\perp\perp}^A$, $R_{\perp\parallel}$. This assumption has the effect that the inclination of the fracture plane given by the fracture plane angle θ_{fp} and the IFF fracture mode connected with it are not affected by the weakening which is now solely dependent on σ_1, since σ_1 does not depend on θ. **Searching the fracture angle can be carried out in the same way as before neglecting an influence of σ_1.**

Reducing the fracture resistances results in an increased stress exposure factor. Therefore the stress exposure factor f_{E_1} when taking into account the influence of σ_1 becomes:

$$f_{E_1} = \frac{f_{E_0}}{\eta_{w1}} \qquad \text{(Eq. 79)}$$

As in [Puck 1996, Puck and Schürmann 2002] the weakening which increases progressively with σ_1 is described by a fracture curve in the form of a segment of an ellipse, Fig. 46. Previously η_{w1} had always to be calculated iteratively but now a closed solution has been found. The section of the ellipse starts at "$s \cdot R_{\|}$" and ends at the FF-limit namely at $|\sigma_1| = R_{\|}$. Here, at the FF-limit the weakening factor η_{w1} reaches its minimum "m". Both parameters "s" and "m" can be chosen independently between 0 and 1 resulting in an excellent adaptability to experiments, especially as the parameter pairs s and m can be set differently in the tension area and in the compression area of σ_1 (see Fig. 46).

Under the given assumptions the following formula for the weakening factor η_{w1} has been derived:

$$\eta_{w1} = \frac{c\left(a\sqrt{c^2(a^2 - s^2) + 1} + s\right)}{(c\,a)^2 + 1} \qquad \text{(Eq. 80)}$$

with $c = \dfrac{f_{E_0}}{f_E(FF)}$ and $a = \dfrac{1-s}{\sqrt{1-m^2}}$.

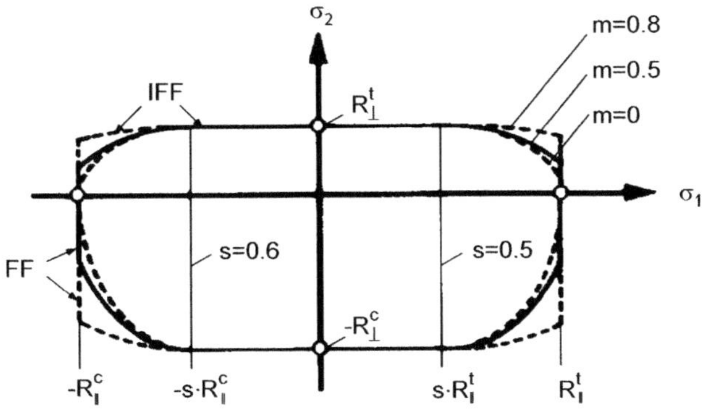

Fig. 46. Form of a (σ_1, σ_2)-fracture curve with influence of σ_1 on the IFF expressed as a function of parameters s and m

$f_E(FF)$ is the FF-stress exposure factor, which is calculated by using the FF-condition $|\sigma_1| = R_{\parallel}$ given in (Eq. 14). The stress exposure factor f_{E_0} is the IFF exposure without an influence of σ_1 (The subscript 0 indicates that $\sigma_1 = 0$). If weakening due to (m+p) effects is not taken into account, then what should be used for f_{E_0} are the stress exposure values f_E calculated with the aid of (Eq. 71), (Eq. 73) and (Eq. 77) or the fracture criteria from Fig. 42 with measured basic strengths and with the guide values given in Table 1, chapter 4.2.6.6 for the inclination parameters. When (m+p) effects are in-cluded, what should be used for f_{E_0} is the value of $f_{E_{m+p}}$ according to (Eq. 88) and using the corrected basic strengths $R_{\perp}{}^{t}{}_{cor}$, $R_{\perp}{}^{c}{}_{cor}$, $R_{\perp\parallel cor}$, and the corrected parameter $p^{c}{}_{\perp\parallel cor}$.

The range of validity of the weakening factor η_{w1} is given by

$$\frac{1}{s} \geq \frac{f_{E_0}}{f_E(FF)} \geq m \qquad \text{(Eq. 81)}$$

Beyond this region there is either no damage or no IFF occurs before the FF-limit is reached. In the later case it makes no sense to talk about an IFF-stress exposure factor. Instead, there is just a FF-stress exposure factor. This is because FF usually results in an extensive destruction of the fiber-matrix-composite. Thus there is no such thing as an independent IFF anymore.

So far there are no reliable experimentally determined values for the parameters s and m available [Kaiser Kuhnel Obst 2004]. Thus assumed values must be chosen. It is recommended to use $s = 0.5$ and $m = 0.5$ both at $\sigma_1 > 0$ and at $\sigma_1 < 0$. With these values η_{w1} becomes

$$\eta_{w1} = \frac{c\left(\sqrt{c^2+12}+3\right)}{2\left(c^2+3\right)} \quad \text{for } 2 \geq c \geq 0.5 . \qquad \text{(Eq. 82)}$$

In principle, one might doubt whether the weakening by η_{w1} could influence the fracture angle θ_{fp}. Thus, the question whether there is such an influence or not is discussed in the following. For doing so, the expression for η_{w1} documented in (Eq. 80) needs to be examined further. Based on the assumption that IFF will occur on the action plane with a maximum of the stress exposure f_{E_1} (including weakening by σ_1), roots of $d(f_{E_1}(\theta))/d\theta$ have to be examined:

$$f_{E_1}(\theta) = \frac{f_{E_0}(\theta)}{\eta_{w1}} = f_{E_0}(\theta)\frac{(ac)^2+1}{c\left(a\sqrt{c^2\left(a^2-s\right)+1}+s\right)} \qquad \text{(Eq. 83)}$$

$$\text{with } c = \frac{f_{E_0}(\theta)}{f_{E(FF)}}$$

Here, $f_{E(FF)}$ does not depend of θ and is therefore not of interest for the search for an extremum.

$$\frac{d\left(f_{E_1}(\theta)\right)}{d\theta} = \frac{d}{dc}\left(f_{E_1}(c)\right)\frac{dc}{d\theta} = \frac{d}{dc}\left(f_{E_1}(c)\right)\frac{df_{E_0}}{d\theta} = 0 \qquad \text{(Eq. 84)}$$

This means, that the stress exposure f_{E_1} including the weakening effect of σ_1 has an extremum at the same point as the stress exposure f_{E_0} calculated without weakening by σ_1. Now the question left to be answered is whether there are further extrema, namely a solution for

$$\frac{d}{dc}\left(f_{E_1}(c)\right) = 0 . \qquad \text{(Eq. 85)}$$

The result for (Eq. 85) is

$$c^2 = -\frac{1}{a} . \qquad \text{(Eq. 86)}$$

This means, that there is no real solution. There are no further extrema of the stress exposure apart from those which are valid for the stress exposure without weakening by σ_1, too.

4.3.1.2 Inclusion of stresses $\sigma_n(\theta), \tau_{nt}(\theta), \tau_{n1}(\theta)$ which act on parallel-to-fiber planes but not on the fracture plane

Physical fundamentals

The effects which are examined will first be presented by taking the example of a (σ_2, τ_{31})-stress combination, where σ_2 is to be a tensile stress. According to Mohr's hypothesis, with (σ_2, τ_{31})-combinations of this kind there either occurs a $\sigma_\perp$-tensile fracture on the action plane of σ_2 – in other words, at $\theta_{fp} = 0°$ when $\sigma_2 = R_\perp^t$ – or a $\tau_{\perp\|}$-shear fracture on the action plane of τ_{31} – in other words, at $\theta_{fp} = \pm 90°$ and this when $\tau_{31} = R_{\perp\|}$. Which fracture actually does occur depends on where the (σ_2, τ_{31})-stress state vector intersects the fracture curve which, according to Mohr's hypothesis, consists of the two straight lines $\sigma_2 = R_\perp^t$ and $\tau_{31} = R_{\perp\|}$ (fracture curve for (σ_2, τ_{31}), see Fig. 47). If we had a uniaxial σ_2-stress state, then a stress $\sigma_n(\theta) = \sigma_2 \cdot \cos^2\theta$ and $\tau_{nt}(\theta) = -\sigma_2 \cdot \sin\theta \cdot \cos\theta$ would act on the sections adjacent to the fracture plane at angles of inclination $\theta \neq 0°$.

We will now consider a (σ_2, τ_{31})-vector which meets the straight line $\sigma_2 = R_\perp^t$ close to its intersection with the straight line $\tau_{31} = R_{\perp\|}$. In this case

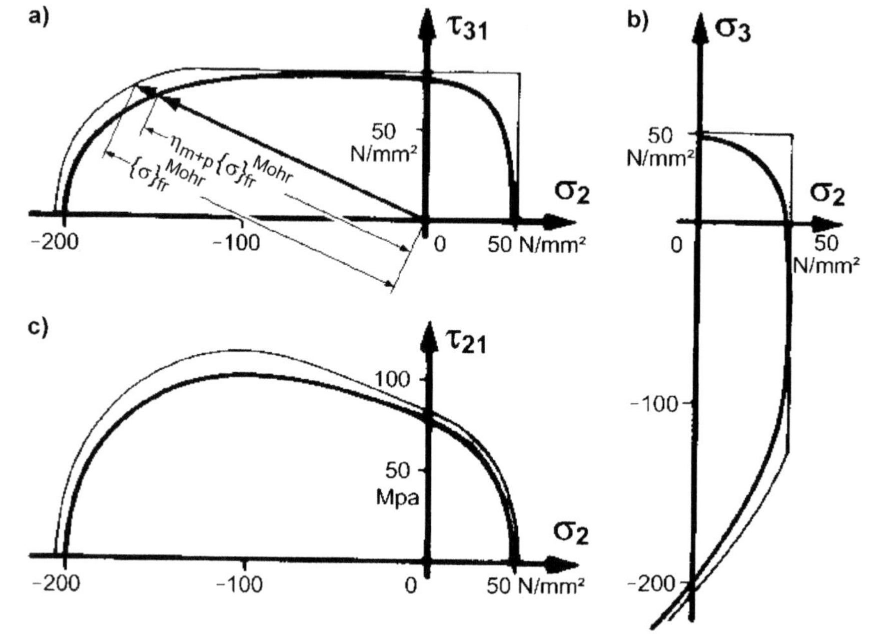

Fig. 47. Fracture curves for three stress combinations: (a) (σ_2, τ_{31}), (b) (σ_2, σ_3) and (c) (σ_2, τ_{21}). The thin lines represent the starting curves for η_{m+p}-correction, calculated using the equations given in Fig. 42 with corrected basic strength values and corrected parameter $p_{\perp\parallel}{}^{c}{}_{cor}$ according to (Eq. 91): $R_{\perp}{}^{t}{}_{cor} = 48.57$ N/mm², $R_{\perp}{}^{c}{}_{cor} = 204.82$ N/mm², $R_{\perp\parallel cor} = 82.16$ N/mm², $p_{\perp\parallel}{}^{t} = 0.35$, $p_{\perp\parallel}{}^{c}{}_{cor} = 0.45$, $p_{\perp\perp}{}^{t} = p_{\perp\perp}{}^{c} = 0.275$. The thick lines represent the result curves of an η_{m+p}-correction according to (Eq. 87), (Eq. 88), (Eq. 89) and (Eq. 90).

the angle of the fracture plane according to Mohr will clearly also be $\theta_{fp} = 0°$. But if stresses σ_2 and τ_{31} are present simultaneously, in addition to $\sigma_n(\theta)$ and $\tau_{nt}(\theta)$ a transverse/longitudinal shear stress $\tau_{n1}(\theta) = \tau_{31} \cdot \sin\theta$ will occur on sections at an angle $\theta \neq 0°$. With this, as $\theta = -90°$ and $\theta = +90°$ is approached and provided $\tau_{31}/R_{\perp\parallel}$ is only a little less than $\sigma_2/R_{\perp}{}^{t}$, there follows an IFF stress exposure $f_E(\theta)$ which is almost as high as that on the action plane at $\theta_{fp} = 0°$ (see standardized stress exposure $f_E'(\theta)$ in Fig. 48). For the combination of σ_2 with τ_{31} which we are considering, the IFF stress exposure over the whole range between $\theta = -90°$ and $\theta = +90°$ will not be much lower than on the action plane itself.

Micromechanical studies show that an IFF does not simply happen suddenly for no particular reason. Once the IFF stress exposure f_E' (standardized to $f_{Emax} = 1$) exceeds a threshold value of about $f_E' \approx 0.5$, instances of

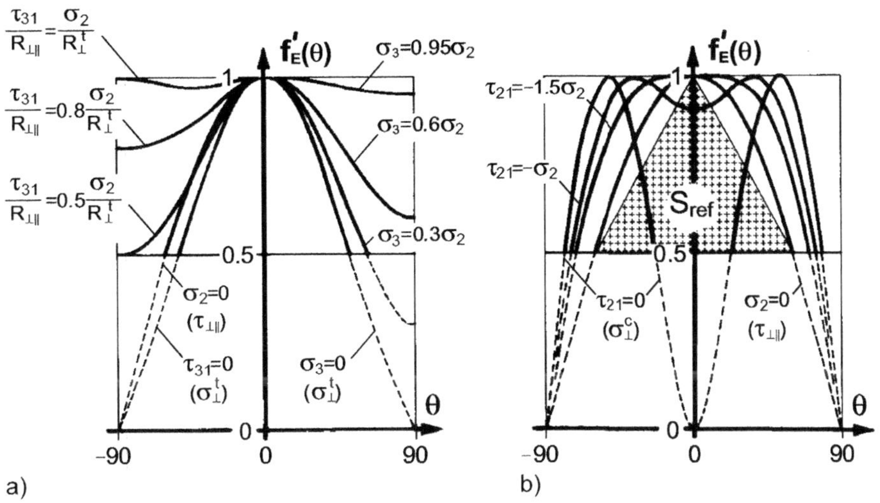

Fig. 48. Standardized stress exposure curves $f_E'(\theta)$ (parameters from Fig. 47); a) Left-hand side of diagram for different (σ_2, τ_{31})-stress combinations; right-hand side for (σ_2, σ_3)-stress combinations. The complete $f_E'(\theta)$-curves for (σ_2, τ_{31})- and (σ_2, σ_3)-stress combinations are symmetrical to the f_E' axis at $\theta = 0°$. b) For different (σ_2, τ_{21})-stress combinations. The triangular area marked in the figure gives an impression of the magnitude of the reference sum S_{ref}

micro-damage to the fiber/matrix composite will already occur which increase progressively as stress increases. In the case of transverse/longitudinal shear stressing $\tau_{\perp\parallel}$, these are the familiar 45° micro cracks (hackles) which are stopped at fibers and cause often tiny delaminations there (see Fig. 49). It is only when micro-fractures have exceeded a certain magnitude that the IFF will suddenly occur.

In the case of the (σ_2, τ_{31})-combination under consideration here, not only micro-fractures resulting from σ_2 but also those resulting from τ_{31} develop simultaneously – the case is different with a uniaxial σ_2 stress. The latter type of micro-fractures due to τ_{31} further weakens the fiber/matrix composite as compared with the situation with uniaxial σ_2-stress. The micro-fractures have a weakening effect on ***all sections*** with inclination angles between $\theta = -90°$ and $\theta = +90°$. ***They also reduce the transverse tensile strength*** which in the end is still available on the fracture plane when IFF happens at $\theta_{fp} = 0°$. It is therefore to be expected that σ_2 in combination with a relatively high τ_{31} stress will not reach at IFF that fracture stress which is obtained in the uniaxial transverse tension test with $\sigma_2 = R_\perp^t$.

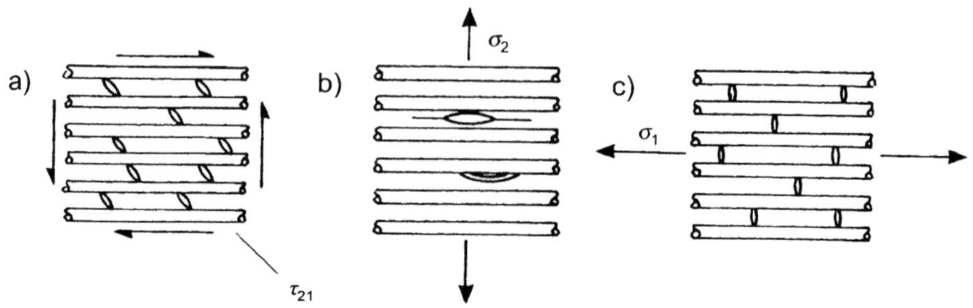

Fig. 49. Occurrence of matrix cracks [Puck 1992]; a) τ_{21} shear stress leads to cracks under 45° (so called hackles) which are stopped at the fibers and are cause of parallel to fiber mini delaminations. b) With σ_2 tensile stress cracks are particularly encouraged where there are flaws and can spread along the fibers without hindrance. c) σ_1 stress results in cracks occurring at the tip of broken individual fibers and running transverse with respect to the fiber until stopped at adjacent fibers.

As a further example let us examine a biaxial (σ_2, σ_3)-transverse tensile stress. In this case a tensile fracture occurs at $\theta_{fp} = 0°$ if $\sigma_2 > \sigma_3$ or at $\theta_{fp} = \pm 90°$ if $\sigma_3 > \sigma_2$. If $\sigma_2 = \sigma_3 = \sigma$ were the case, the same stress would prevail at all sections at any angle θ due to the relation $\sigma(\theta) = \sigma_2 \cdot \cos^2\theta + \sigma_3 \cdot \sin^2\theta = \sigma$ and thus the intersection-angle-dependent stress exposure $f_E(\theta)$ would be equally high on every section plane with θ between −90° and +90°. In this special individual case Mohr's hypothesis predicts that IFF will occur simultaneously on all section planes with angles of θ between −90° and +90°. At what angle fracture will actually take place in a biaxial transverse tension test with two stresses of equal magnitude depends on contingencies, e. g. on the distribution of flaws.

We will now pass on to a (σ_2, σ_3)-combination where $\sigma_3 = 0.95 \cdot \sigma_2$. In this case the stress exposure $f_E(\theta)$ is nearly the same in all sections with angles θ between −90° and +90° (see again the standardized stress exposures $f'_E(\theta)$ in Fig. 48). For this reason a more massive instance of microdamage is to be expected before the IFF occurs than would be the case with uniaxial transverse tensile stress.

A real fiber/matrix composite in addition always contains flaws – for example, in the form of curing cracks, flat air entrapments, or local imperfections in the bonding between fiber and matrix which often cover only a part of the circumference of the fibers. Flaws of this kind have a 'sense of direction'. For example, a curing crack will have an especially strength-reducing effect when a tensile stress σ_n perpendicular to the crack plane occurs. In the case under consideration, in which σ_3 is only 5% less than

σ_2, the stress exposure $f_E(\theta)$ is theoretically at its maximum at $\theta_{fp} = 0°$. There is nevertheless a high probability that with a different angle of inclination θ of the section plane, a particularly serious flaw will trigger the IFF despite the somewhat lower theoretical stress exposure $f_E(\theta)$ found there. This is a probabilistic effect. It makes prediction of the angle of the fracture plane uncertain and causes the fracture stresses at IFF to be somewhat lower than those calculated on the basis of Mohr's hypothesis.

In both examples there is a mixture of the effects of microdamage and of probabilistic effects; they cannot be treated separately. Generalizing, these examples permit us to arrive at the following conclusion. ***It is to be expected that the effects of microdamage and probabilistics reduce the magnitude of fracture stresses when the IFF occurs will rise the more parallel-to-fiber sections at different angles θ there are for which relatively high values of the stress exposure factor $f_E(\theta)$ can be calculated*** – or, to put it another way, the 'rounder' the 'stress exposure curve' $f_E(\theta)$ will be. A clear picture of which situation of this kind exists with the stress state to be investigated can be obtained by examining the entire stress exposure curve $f_E'(\theta)$ calculated using Mohr's equations of Fig. 42 (see standardized stress exposures in Fig. 48). The stress combination $\sigma_2 = \sigma_3 > 0$ has been recognized as being an extreme case where at any section the value for $f_E'(\theta)$ is equally high.

Analytical treatment

Micromechanical failure analyses and mathematical methods from probabilistics cannot be used in component design. A calculation method will therefore be presented below whereby it is possible, with minimal effort, to estimate the effects of **m**icro-damage and **p**robabilistics (in abbreviated form: (m+p) effects) with the aid of a phenomenological approach on physical foundations and which is applicable in engineering practice. This method will be referred to as η_{m+p} **correction**. It is a slightly modified form of the approach which appears in [Puck 1996]. The η_{m+p}-correction is based on the following assumptions:

The fracture plane at IFF appears at the angle of intersection θ_{fp} for which, according to Mohr's hypothesis the maximum stress exposure $f_E(\theta)_{max} = f_E(\theta_{fp})$ is calculated from the fracture criteria in Fig. 42.

The fracture stresses at IFF are obtained by correcting the fracture stresses calculated from $f_E(\theta_{fp})$ by Mohr's hypothesis to lower stresses using a correction factor of $\eta_{m+p} < 1$. (Correspondingly, the IFF stress exposures calculated by Mohr are corrected to higher values by dividing the Mohr stress exposures by η_{m+p}).

Expressed formally (see also Fig. 47a):

$$\{\sigma\}_{m+p} = \{\sigma\}_{fr}^{Mohr} \cdot \eta_{m+p} = \frac{\{\sigma\}}{f_E(\theta_{fp})} \cdot \eta_{m+p} = \frac{\{\sigma\}}{f_{E_{m+p}}} \qquad \text{(Eq. 87)}$$

$$f_{E_{m+p}} = \frac{f_E(\theta_{fp})}{\eta_{m+p}}. \qquad \text{(Eq. 88)}$$

$\{\sigma\}_{m+p}$ = fracture stress vector taking (m+p) effects into account,
$\{\sigma\}_{fr}^{Mohr}$ = fracture stress vector without taking (m+p) effects into account,
$\{\sigma\}$ = vector of the effective stress state,
η_{m+p} = correction factor of the fracture stresses for taking (m+p) effects into account,
$f_{E_{m+p}}$ = stress exposure of the fracture plane taking (m+p) effects into account,
$f_E(\theta_{fp})$ = stress exposure of the fracture plane without taking (m+p)-effects into account.

As will be explained later, $\{\sigma\}_{fr}^{Mohr}$ and $f_E(\theta_{fp})$ have to be calculated by using corrected strengths $R_{\perp}^{t}{}_{cor}$, $R_{\perp}^{c}{}_{cor}$, $R_{\perp\|cor}$. The reason for this is the following. Experimentally determined strengths values are "real" values. This means that effects of micro-damage and probabilistics have already influenced these values. The analytical procedure of the η_{m+p}-correction should therefore not a second time degrade these values. In other words: When operating the η_{m+p}-procedure and applying the η_{m+p}-correction to the case of uniaxial transverse tension, to the case of uniaxial transverse compression and to pure longitudinal shear the results should be the experimental values for $R_{\perp}^{t}$, $R_{\perp}^{c}$ and $R_{\perp\|}$. This is achieved in the following way:

Before starting a fracture analysis with η_{m+p}-correction the experimentally determined values $R_{\perp}^{t}$, $R_{\perp}^{c}$ and $R_{\perp\|}$ are divided by η_{m+p} belonging to uniaxial tensile stressing, uniaxial compressive stressing and pure longitudinal shear stressing respectively. The appropriate factors η_{m+p} for correcting $R_{\perp}^{t}$, $R_{\perp}^{c}$ and $R_{\perp\|}$ are calculated from (Eq. 89) and (Eq. 90) by using the corrected strengths $R_{\perp}^{t}{}_{cor}$, $R_{\perp}^{c}{}_{cor}$ and $R_{\perp\|cor}$. These are not known at the beginning. This is no problem as far as $R_{\perp}^{c}{}_{cor}$ and $R_{\perp\|cor}$ are concerned, because when normalizing f_E to f_E' the chosen values for these strengths are cancelled from the equation. This is different for uniaxial tension, where f_E is dependent on $R_{\perp}^{t}{}_{cor}$ and $R_{\perp\perp}^{A}{}_{cor}$. In this case a few iterations are necessary.

Later on, perhaps also the sustainable strengths for uniaxial $\sigma_{\perp}^{t}$ and $\sigma_{\perp}^{c}$ or pure $\tau_{\perp\|}$ are calculated in the course of a general fracture analysis including (m+p) effects. This is done now automatically by multiplying the corrected strengths $R_{\perp}{}^{t}{}_{cor}$, $R_{\perp}{}^{c}{}_{cor}$ and $R_{\perp\|cor}$.with just those η_{m+p} by which $R_{\perp}{}^{t}$, $R_{\perp}{}^{c}$ and $R_{\perp\|}$ had been divided before. Therefore, the results of these η_{m+p}-corrections are just the experimentally determined strengths $R_{\perp}{}^{t}$, $R_{\perp}{}^{c}$ and $R_{\perp\|}$.

The dependence of the (m+p) correction factor η_{m+p} on the "roundness" of the standardized stress exposure curve $f_E'(\theta)$ can be quantified by means of the following evaluation of the stress exposure curve $f_E'(\theta)$with the aid of a "summation" formula. In the fracture analysis of a general three-dimensional stress state, during the necessary numerical search for the fracture plane angle the intersection-angle-dependent IFF stress exposure $f_E(\theta)$ from $\theta = -90°$ to $\theta = +90°$ is calculated with the aid of the fracture criteria in Fig. 42 in small angular increments $\Delta\theta$ (in most cases with $\Delta\theta = 1°$). The result is a large number of $f_E(\theta)$-values which represent points of the stress exposure curve $f_E(\theta)$. To ensure that all possible stress states are treated alike, the stress exposure curves are standardized in such a way that at point $\theta = \theta_{fp}$ the stress exposure factor has a value of 1. The standardized stress exposure curve $f_E'(\theta)$ is then valid for just that magnitude of the stresses which according to the fracture criteria in Fig. 42 would produce fracture and this is true for any kind of stress combination.

It follows from the foregoing that section planes with $f_E'(\theta)$-values close to 1 are of much greater importance than those with $f_E'(\theta)$ values which are considerably smaller. For this reason, as already occurs in [Puck 1996], standardized $f_E'(\theta)$ values below a threshold $f'_{E_{thr}} = 0.5$ are not taken into consideration. On the basis of equation (10.5) in [Puck 1996], the following sum is obtained with the standardized $f_E'(\theta)$ values:

$$S \approx \sum_{-90°}^{+89°} \left(f'_E(\theta) - f'_{E_{thr}} \right) \cdot \Delta\theta \text{, only for } f'_E(\theta) \geq f'_{E_{thr}} . \qquad \text{(Eq. 89)}$$

(The limits –90° and +89° apply to the angular increment $\Delta\theta = 1°$ which was used).

This summing (with a zero line shifted by $f'_{E_{thr}}$) results in high $f_E'(\theta)$ values being weighted considerably more heavily than low values – for example, a value $f_E'(\theta) = 0.9$ is weighted four times higher than a value $f_E'(\theta) = 0.6$. For the extreme case which we mentioned of $\sigma_2 = \sigma_3 > 0$, when $f_E'(\theta) = 0.5$ the maximum S value of $S_{max} = (1–0.5)\cdot180° = 90°$ is obtained.

If the summation formula (Eq. 89) is applied to the standardized stress exposure curves $f_E'(\theta)$ for the basic stressings, uniaxial transverse tension $\sigma_{\perp}{}^{t}$, uniaxial transverse compression $\sigma_{\perp}{}^{c}$ and pure transverse/longitudinal shear $\tau_{\perp\|}$ (see Fig. 17), we obtain values for the corresponding sums $S_{\perp}{}^{t}$,

$S_\perp^c$, $S_{\perp\parallel}$ which will lie between 30° and 40°. $S_{\perp\parallel} = 39.24°$ is a material-independent fixed value while $S_\perp^c$ depends on the inclination parameter $p_{\perp\perp}^c$ and has a value of approximately $S_\perp^c \approx 36°$. For $S_\perp^t$ we find material dependent values with $S_\perp^t \approx 33°$.

Taking the value S which corresponds to a given stress state and slightly modifying equation (10.6) in [Puck 1996], we obtain the following for the (m+p)-correction factor:

$$\eta_{m+p} = 1 - \Delta_{max} \frac{S - S_{ref}}{S_{max} - S_{ref}} \qquad \text{(Eq. 90)}$$

for $\Delta_{max} \leq 0.25$ with $S_{max} = 90°$ and $S_{ref} = 30°$.

It is necessary to introduce a reference value S_{ref} in order to obtain a reasonable sensitivity of η_{m+p} to the differences in the S values occurring with different stress states. Experience shows that there is no S value of any stress state which is less than the value $S_{ref} = 30°$. The variable Δ_{max} is the relative difference between the fracture stress vector calculated according to Mohr and that corrected by η_{m+p} when $\sigma_2 = \sigma_3 > 0$. Values for Δ_{max} should lie between 0.15 and 0.25. Selection of a value for Δ_{max} results in a 'calibration' of the η_{m+p}-correction factor.

Applications

In common practice of the design and dimensioning of fiber composite components, fracture criteria are used for calculating the IFF stress exposure (and FF stress exposure) at some locations of the component which are regarded as critical. In what follows, however, entire fracture curves for IFF will be presented. Fracture curves of this kind reveal in which stress states η_{m+p}-correction is important and in which it is not. In Fig. 47a it can be seen that with the (σ_2, τ_{31})-combinations, η_{m+p}-correction is very important in the first quadrant. With the (σ_2, σ_3)-combinations (Fig. 47b), a very noticeable influence also occurs in the first quadrant of the (σ_2, σ_3)-fracture curve. In the case where $\sigma_2 = \sigma_3$, this yields the utilized "calibration value" $\Delta_{max} = 0.25$ used with η_{m+p}-correction. In the fourth (and second) quadrants, on the other hand – that is, where $\sigma_2 > 0$ and $\sigma_3 < 0$ – only a relatively minor effect of the η_{m+p}-correction is found. At $\sigma_3 = -\sigma_2$ the fracture stress at IFF calculated with η_{m+p} is even a little higher than that calculated on the basis of Mohr's hypothesis ($\sigma_2 = -\sigma_3 = R_\perp^t$). This results from the fact that for the pure transverse/transverse shearing stressing $\tau_{\perp\perp}$ which is here present the corresponding sum has the value $S_{\perp\perp} = 31.45°$, and this is somewhat less than $S_\perp^t = 32.45°$ for uniaxial transverse tension stressing $\sigma_\perp^t$.

To date there has been a lack of credible experimental results for (σ_2, τ_{31})- and (σ_2, σ_3)-stress combinations.

The situation is entirely different for the (σ_2, τ_{21})-stress combinations shown in Fig. 47c. The (σ_2, τ_{21})-fracture curve is the only one which has adequate experimental backing not only for CFRP but also for GFRP [Cuntze et al. 1997]. An outstandingly good mathematical model is obtained with equations (Eq. 71), (Eq. 73) and (Eq. 77) or the fracture criteria in Fig. 42 using inclination parameters from Table 1, chapter 4.2.6.6. Before proceeding further, it is important to bear in mind the fact that the effects of η_{m+p}-effects on IFF strength are basically already included in experimentally determined fracture stresses, as also in the test results from which the modeled (σ_2, τ_{21})-fracture curve (thick line in Fig. 47c) is obtained. If one wished to generate mathematically a (σ_2, τ_{21})-fracture curve which involved use of η_{m+p}-correction, then the experimentally obtained (σ_2, τ_{21})-fracture curve which was modeled using the measured basic strengths $R_\perp^t$, $R_\perp^c$, $R_{\perp\parallel}$ and inclination parameters from Table 1 is not the *starting curve* for the η_{m+p}-correction procedure but represents rather the *result curve.*

If it is exclusively plane $(\sigma_1, \sigma_2, \tau_{21})$-stress states which have to be investigated, then the question of the unknown starting curve for an η_{m+p}-correction is irrelevant since the mathematically modeled, experimentally determined (σ_2, τ_{21})-fracture curve is being used directly for fracture analysis. On the other hand, use of η_{m+p}-correction is advisable in those cases where it is necessary, during dimensioning of FRP components, to also analyze load application zones in which spatial stresses σ_3, τ_{32}, τ_{31} of a similar magnitude to σ_2 and τ_{21} occur. To ensure that absolutely consistent calculation results are obtained here, it is, of course, necessary that the same parameters be used uniformly in the fracture criteria in Fig. 42 for all stress states which occur – in other words, even for the $(\sigma_1, \sigma_2, \tau_{21})$-stress states prevailing in undisturbed areas. For this reason the parameters must be known which are appropriate for the (σ_2, τ_{21})-starting curve which is itself as yet unknown.

Since in the case of the (σ_2, τ_{21})-stress combinations the *result curve* is known and the starting curve is required, an inversion of the η_{m+p}-correction, as it were, is now necessary. The corrected basic strengths $R_{\perp\,cor}^t$, $R_{\perp\,cor}^c$, $R_{\perp\parallel cor}$ for the starting curve are obtained by 'inverting' the η_{m+p}-correction: the measured values $R_\perp^t$, $R_\perp^c$, $R_{\perp\parallel}$ are not multiplied but are instead now divided by the η_{m+p}-value which is associated with the corresponding basic stressing. (Comment: These corrected basic strengths of course apply not only to the (σ_2, τ_{21})-stress state but of course also to any stress state.) Since the sum values $S_\perp^t$, $S_\perp^c$, $S_{\perp\parallel}$ of the basic stressings are only a little higher than the reference value $S_{ref} = 30°$, the correction they need remains less than 5%. In the range of high compressive stress values $|\sigma_2|$ with at the same time high values of τ_{21}, sum values up to around $S \approx 60°$ are however reached in the vicinity of the transition point between Mode B ($\theta_{fp} = 0°$) and Mode C ($\theta_{fp} \neq 0°$) (see Fig. 24). This results in η_{m+p}-corrections up to approximately 12%, see Fig. 48b and Fig. 47c.

The required starting curve of the (σ_2, τ_{21})-stress combination cannot be obtained as a whole from the known result curve simply by inverting the η_{m+p}-correction. The reason for this lies in the complex nature of the η_{m+p}-correction which is only to be carried out numerically on a point by point basis and which makes impossible a closed mathematical formulation of the required starting curve. What is needed, however, is a (σ_2, τ_{21}) starting curve formulated using the equations in Fig. 42, since only these equations happen to be available for the fracture analysis of general three-dimensional stress states. For this reason, an acceptable approximated starting curve which is described using the equations in Fig. 42 must be found by selecting suitable parameters. What acceptable means is that when a η_{m+p}-correction of this starting curve has been carried out correctly, a (σ_2, τ_{21})-fracture curve is obtained as the result which will deviate only to an acceptable extent from the known modeled experimental (σ_2, τ_{21})-fracture curve.

By iterative selection of different inclination parameters $p_{\perp\parallel}^{c}{}_{cor}$ for the (σ_2, τ_{21})-starting curve and reviewing to what extent the required acceptance is thereby achieved, the following result is obtained. If a corrected inclination parameter $p_{\perp\parallel}^{c}{}_{cor} = 0.45$ is chosen for the required starting curve of the η_{m+p} correction when the calibration value $\Delta_{max} = 0.25$ is used, what is obtained as the result of a correctly performed η_{m+p}-correction is a (σ_2, τ_{21})-fracture curve which comes very close to the experimental curve modeled in the usual way using the equations in Fig. 42. (Modeling of the experimental curve was carried out using the basic strengths for CFRP $R_{\perp}^{t} = 48$ N/mm^2, $R_{\perp}^{c} = 200$ N/mm^2, $R_{\perp\parallel} = 79$ N/mm^2 [Soden et al. 1998] and the inclination parameters $p_{\perp\parallel}^{t} = 0.35$, $p_{\perp\parallel}^{c} = 0.30$, $p_{\perp\perp}^{t} = p_{\perp\perp}^{c} = 0.275$ according to [Puck Kopp Knops 2002]).

If, in response to one's own experimental results, one wishes to use a value different from the value $\Delta_{max} = 0.25$ used in the example and also, within the permitted limits [Puck Kopp Knops 2002], to diverge from the guideline value $p_{\perp\parallel}^{c} = 0.3$ given in Table 1, you can proceed iteratively as described above to obtain a corrected parameter $p_{\perp\parallel}^{c}{}_{cor}$ provided you have a computer program which can display (σ_2, τ_{21})-fracture curves. If this is not the case, the following extrapolation formula may be used within a restricted range of $p_{\perp\parallel}^c$ and Δ_{max}:

$$\frac{p^c_{\perp\parallel cor}}{p^c_{\perp\parallel}} = 1 + 0.6 \cdot \frac{\Delta_{max}}{p^c_{\perp\parallel}} \qquad \text{(Eq. 91)}$$

for $0.15 \leq \Delta_{max} \leq 0.25$ and $0.25 \leq p^c_{\perp\parallel} \leq 0.35$.

Comment

It has been shown that depending on whether you work with or without η_{m+p}-correction, you will have to analyze the stress exposure curve $f_E(\theta)$ by using different inclination parameters $p_{\perp\parallel}{}^c{}_{cor}$ or $p_{\perp\parallel}{}^c$ respectively which may differ according to (Eq. 91) by a factor of up to 1.6. If shear stresses τ_{21} and/or τ_{31} occur in the stress state under investigation, by calculating with and without η_{m+p}-correction you will therefore obtain angles of the fracture plane θ_{fp} which differ somewhat. In the case of (σ_2, τ_{31})-stress combinations the differences are from 1° to 5°. With (σ_2, τ_{21})-stress combinations similar differences are found, provided the angles of the fracture plane $|\theta_{fp}| \geq 30°$. Considerably greater differences occur in the vicinity of the transition point from mode B to mode C (see Fig. 24). But here any statements about the angle of the fracture plane which actually occurs will be tainted with a high degree of uncertainty anyway: this is due to the probabilistic effects which occur and which can be expected due to the flat course of the standardized stress exposure curve. This has been shown in experiments as well [Kopp 2000]. All in all, these deviations are not problematic in the fracture analysis.

4.3.2 Calculation of the stretch factor f_S^L of the load-determined stresses when residual stresses are present

4.3.2.1 Basic considerations

In the following the procedure to be used when calculating the stretch factor $f_S{}^L$ of the (changing) load-determined lamina stresses $\sigma_1{}^L$, $\sigma_2{}^L$, $\tau_{21}{}^L$ when (constant) residual stresses $\sigma_1{}^r$, $\sigma_2{}^r$, $\tau_{21}{}^r$ are simultaneously present will be explained for the case of a (σ_1, σ_2, τ_{21}) stress state. This procedure is based on Puck's action-plane fracture criteria for IFF and takes into consideration the weakening influence of the parallel-to-fiber stress σ_1. The formulae required for the procedure are given.

The fracture conditions are the starting point for IFF when there is no influence from σ_1. These are (see equations (Eq. 71), (Eq. 73) and (Eq. 77)):

IFF mode A

$$f_{E_0} = \sqrt{\left[\left(\frac{1}{R_\perp^t} - \frac{p_{\perp\parallel}^t}{R_{\perp\parallel}}\right) \cdot \sigma_2\right]^2 + \left(\frac{\tau_{21}}{R_{\perp\parallel}}\right)^2} + \frac{p_{\perp\parallel}^t}{R_{\perp\parallel}} \sigma_2 = 1 \text{ , } for\ \sigma_2 \geq 0 \qquad \text{(Eq. 71)}$$

IFF mode B

$$f_{E_0} = \sqrt{\left(\frac{\tau_{21}}{R_{\perp\parallel}}\right)^2 + \left(\frac{p_{\perp\parallel}^c}{R_{\perp\parallel}} \sigma_2\right)^2} + \frac{p_{\perp\parallel}^c}{R_{\perp\parallel}} \sigma_2 = 1, \sigma_2 < 0 \wedge \left|\frac{\sigma_2}{\tau_{21}}\right| \leq \left|\frac{R_{\perp\perp}^A}{\tau_{21c}}\right| \qquad \text{(Eq. 73)}$$

IFF mode C

$$f_{E_0} = \frac{\tau_{21}^2}{4\left(R_{\perp\parallel} + p_{\perp\parallel}^c R_{\perp\perp}^A\right)^2} \cdot \frac{\left(-R_\perp^c\right)}{\sigma_2} + \frac{\sigma_2}{\left(-R_\perp^c\right)} = 1$$

$$\text{for } \sigma_2 < 0 \text{ and } 0 \leq \left|\frac{\tau_{21}}{\sigma_2}\right| \leq \frac{\left|\tau_{21_c}\right|}{R_{\perp\perp}^A} \qquad \text{(Eq. 92)}$$

(Eq. 92) is equivalent to (Eq. 77). The following coupling of inclination parameters was used in deriving equation (Eq. 92):

$$\frac{p_{\perp\perp}^c}{R_{\perp\perp}^A} = \frac{p_{\perp\parallel}^c}{R_{\perp\parallel}} \qquad \text{(Eq. 93)}$$

For this reason $R_{\perp\perp}^A$ needs to be calculated using equation (Eq. 94):

$$R_{\perp\perp}^A = \frac{R_{\perp\parallel}}{2p_{\perp\parallel}^c} \cdot \left(\sqrt{1 + 2p_{\perp\parallel}^c \frac{R_\perp^c}{R_{\perp\parallel}}} - 1\right) \qquad \text{(Eq. 94)}$$

The fracture conditions for IFF – equations (Eq. 71), (Eq. 73) and (Eq. 77) ((Eq. 92) respectively) – describe in the (σ_1, σ_2, τ_{21}) stress space the circumferential surface of a 'cylindrical' fracture body (see Fig. 50) whose constant cross-sectional contour is the fracture curve for (σ_2, τ_{21}) stress combinations shown in Fig. 45 which has been completed symmetrically with respect to the σ_2 axis. This body extends to infinity in both the positive and negative σ_1 directions; its real importance ceases, however, no later than at the two fiber fracture limits at $\sigma_1 = R_\perp^t$ and $\sigma_1 = -R_\perp^c$.

Any influence of σ_1 on the IFF is taken into account by introducing a weakening factor $\eta_{w1} < 1$ (see above) in the case of the strengths $R_\perp^t$, $R_\perp^c$,

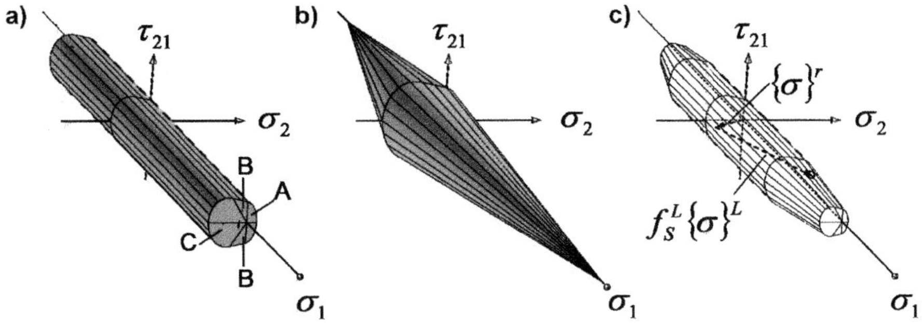

Fig. 50. Creation of the (σ_1, σ_2, τ_{21}) fracture body for IFF. a) (σ_1, σ_2, τ_{21}) fracture body for IFF without σ_1 influence according to (Eq. 71), (Eq. 73) and (Eq. 92). This fracture body which theoretically extends to infinity in both the positive and negative σ_1 direction is of practical relevance only between the fiber fracture limits at $\sigma_1 = R_{\parallel}^t$ and $\sigma_1 = (-R_{\parallel}^c)$. b) IFF fracture body calculated with weakening σ_1 influence using (Eq. 101), (Eq. 102) and (Eq. 103). c) Valid remaining fracture body as intersection of the bodies in accordance with a) and b) and following removal of the invalid parts

$R_{\perp\parallel}$ which are of primary, decisive importance to the IFF. These strengths do not weaken until a threshold value of σ_1 is passed. This value must be determined on the basis of experimental experience and is given as a fraction of the stress $(\sigma_1)_{fr} = R_{\parallel}^t$ or $(\sigma_1)_{fr} = (-R_{\parallel}^c)$ leading to fiber fracture. In chapter 4.3.1.1 an ellipse has been selected for the course of the weakening factor η_{w1} as a function $\eta_{w1}(\sigma_1)$. In order to obtain a closed solution for f_S^L, and that in the form of the solution of a quadratic equation, it will be necessary to linearize the function for η_{w1} (see equation (Eq. 95) and Fig. 51). Weakening of the strengths $R_{\perp}^t$, $R_{\perp}^c$, $R_{\perp\parallel}$ increases as FF stress exposure $f_{E,FF}$ increases; in other words, the weakening factor η_{w1} becomes smaller as $f_{E,FF}$ increases. Here the following linear equation is used:

$$\eta_{w1} = \eta_{w1_0} - \eta_{w1}' \cdot f_{E,FF} \tag{Eq. 95}$$

$\eta_{w1,0}$ is a hypothetical value of η_{w1} when $\sigma_1 = 0$, and η_{w1}' is the slope of the straight line for η_{w1} according to equation (Eq. 95).

By definition, the stress exposure $f_{E,FF}$ is the ratio of the acting stress and the stress leading to fracture (cf. discussion). Accordingly, from the FF condition equation (Eq. 15) we obtain:

$$f_{E,FF} = \frac{\sigma_1}{(\sigma_1)_{fr}} \quad \text{with} \quad \begin{cases} (\sigma_1)_{fr} = R_{\parallel}^t & for\ \sigma_1 \geq 0 \\ (\sigma_1)_{fr} = -R_{\parallel}^c & for\ \sigma_1 < 0 \end{cases} \tag{Eq. 96}$$

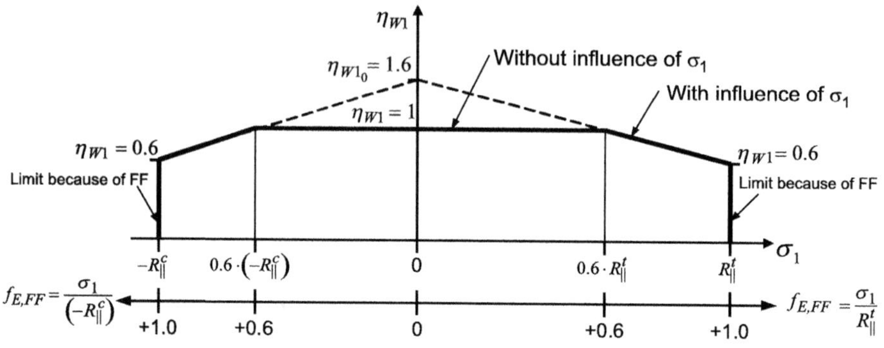

Fig. 51. Course of the weakening factor η_{w1} as a function of σ_1 or as a function of $f_{E,FF}$ in the range $-R_{\parallel}^{c} \geq \sigma_1 \leq R_{\parallel}^{t}$ or $0 < f_{E,FF} < 1.0$ respectively. In order to approximate the ellipse with parameters $s = m = 0.5$ as recommended in chapter 4.3.1.1, the values $\eta_{w1_0} = 1.6$ and $\eta'_{w1} = 1$ are chosen ($S = M = 0.6$).

The reason for using different signs with $R_{\parallel}^{t}$ and $(-R_{\parallel}^{c})$ lies in the international convention whereby compressive stresses are expressed as negative values while all strengths (even compressive strengths) are expressed as positive values.

Now, with the aid of the weakening factor $\eta_{w1} = \eta_{w1}(\sigma_1)$ which we introduced in the case of the three strengths $R_{\perp}^{t}$, $R_{\perp}^{c}$, $R_{\perp\parallel}$ a description is obtained of that (conical) part of the fracture body whose surface equation depends not only on σ_2 and τ_{21} but also on σ_1 (Fig. 50b).

To do so, it is necessary to select numerical values for the two parameters η_{w10} and η'_{w1} in the linear equation (Eq. 95) for the weakening factor η_{w1}. In chapter 4.3.1.1 the "starting point" of the "elliptical" weakening was defined by the value *s* and the minimum value of the weakening factor at the FF limit by the value *m*. If the corresponding values for the linearized course of $\eta_{w1}(f_{E,FF})$ are designed by S and M this delivers the following relationships:

$$\eta_{w1_0} = \frac{1 - S \cdot M}{1 - S} \qquad \text{(Eq. 97)}$$

$$\eta'_{w1} = \frac{1 - M}{1 - S} \qquad \text{(Eq. 98)}$$

In chapter 4.3.1.1 the recommendation was made, should experimental results be lacking, that the values $s = 0.5$ and $m = 0.5$ should be selected when η_{w1} has an elliptical course. A good approximation to this ellipse may be obtained by using $S = 0.6$ and $M = 0.6$ for the straight line. On the basis of (Eq. 97) this yields $\eta_{w10} = 1.6$ and on the basis of (Eq. 98) $\eta_{w1}' = 1.0$. From this we obtain the course of the weakening factor η_{w1} shown in Fig. 51 as a function of σ_1 or $f_{E,\,FF}$. (In principle even different pairs of values (S and M) can be used for the ranges $\sigma_1 > 0$ and $\sigma_1 < 0$.)

By applying an equally high weakening factor η_{w1} to all three strengths $R_\perp^t$, $R_\perp^c$, $R_{\perp\parallel}$ which have a decisive influence on IFF we obtain (σ_2, τ_{21}) fracture curves which represent a geometrically similar reduction by a factor $\eta_{w1} < 1$ of the (σ_2, τ_{21}) fracture curve shown in Fig. 45 which was completed symmetrically to the σ_2 axis. As a consequence, therefore, of the linear equation for $\eta_{w1}(\sigma_1)$, in the (σ_1, σ_2, τ_{21}) stress space we find a fracture body for IFF which has the shape of a double cone (Fig. 50b). Its cross sections perpendicular to the σ_1 axis are geometrically similar.

The "cylindrical" fracture body shown in Fig. 50a and the double-conical fracture body intersect each other at $\sigma_1 = S \cdot R_\parallel^t$ and $\sigma_1 = S \cdot (-R_\parallel^c)$ in two sections which are vertical with respect to the σ_1 axis. The validity of the cylindrical fracture body extends from the coordinates origin as far as these intersection planes. The conical circumferential surfaces of the double-conical fracture body for IFF as shown in Fig. 50b are valid in the range between the intersection planes and the planes at $\sigma_1 = R_\parallel^t$ and $\sigma_1 = -R_\parallel^c$ which are vertical to the σ_1 axis and assigned to FF.

Calculation of the desired stretch factor f_S^L for the load-determined stresses σ_1^L, σ_2^L, τ_{21}^L – from the geometrical point of view – requires determination of the point where the resulting stress vector $\{\sigma\}$ makes contact with the valid range of the surface of the (σ_1, σ_2, τ_{21}) fracture body for IFF (Fig. 50c). The resulting stress vector $\{\sigma\}$ comes from the "geometrical sum" of the residual stress vector $\{\sigma\}^r$ and the load stress vector $\{\sigma\}^L$ elongated (stretched) by the factor f_S^L. The shape of the (σ_1, σ_2, τ_{21}) fracture body as shown in Fig. 50c and the requirement that the load-determined vector $\{\sigma\}^L$ may only be stretched in its specified direction means that there is only one possible point of contact between the vector and the surface of the fracture body. In most cases this point of contact will be located on the cylindrical or conical part of the circumferential surface applicable to IFF. It can, however, also be located on one of the two "end faces" which are valid for FF.

4.3.2.2 Derivation of the formulae required for calculating f_S^L

In the fracture conditions for IFF – equations (Eq. 71), (Eq. 73) and (Eq. 92) – the same weakening factor $\eta_{w1}(\sigma_1)$ is used with the strengths

$R_\perp^t$, $R_\perp^c$, $R_{\perp\parallel}$. Using IFF stress exposure f_{E_1} with weakening due to σ_1 we therefore have as *fracture condition* with weakening:

$$f_{E_1}(\sigma_2,\tau_{21},\sigma_1)=\frac{f_{E_0}(\sigma_2,\tau_{21})}{\eta_{W1}(\sigma_1)}=1 \qquad \text{(Eq. 99)}$$

or also $f_{E_0}(\sigma_2,\tau_{21})=\eta_{W1}(\sigma_1)$ (Eq. 100)

For the three IFF modes this therefore yields:

For mode A

$$\frac{1}{R_{\perp\parallel}}\left[\sqrt{\left(\frac{R_{\perp\parallel}}{R_\perp^t}-p_{\perp\parallel}^t\right)^2\sigma_2^2+\tau_{21}^2}+p_{\perp\parallel}^t\sigma_2\right]=\eta_{W1}(\sigma_1) \qquad \text{(Eq. 101)}$$

For mode B

$$\frac{1}{R_{\perp\parallel}}\left(\sqrt{\tau_{21}^2+\left(p_{\perp\parallel}^c\sigma_2\right)^2}+p_{\perp\parallel}^c\sigma_2\right)=\eta_{W1}(\sigma_1) \qquad \text{(Eq. 102)}$$

For mode C

$$\frac{\tau_{21}^2}{4(R_{\perp\parallel}+p_{\perp\parallel}^c R_{\perp\perp}^A)^2}\cdot\frac{\left(-R_\perp^c\right)}{\sigma_2}+\frac{\sigma_2}{\left(-R_\perp^c\right)}=\eta_{W1}(\sigma_1) \qquad \text{(Eq. 103)}$$

with $\eta_{W1}=\eta_{W1_0}-\eta_{W1}^{'}\cdot\dfrac{\sigma_1}{(\sigma_1)_{fr}}$,

where $(\sigma_1)_{fr}=R_\parallel^t$ for $\sigma_1>0$

and $(\sigma_1)_{fr}=-R_\parallel^c$ for $\sigma_1<0$.

The validity ranges of fracture conditions (Eq. 101) to (Eq. 103) are the same as those for equations (Eq. 71), (Eq. 73) and (Eq. 77) (see also Table 3). In the equations (Eq. 101) to (Eq. 103) the stresses σ_2 and τ_{21} are the *stresses at the fracture limit*.

The strengths $R_\perp^t$, $R_\perp^c$, $R_{\perp\parallel}$ appearing in equations (Eq. 101) to (Eq. 103) are the "unweakened" strengths. For the stresses (σ_1, σ_2, τ_{21}) occurring in them we use the stress components corresponding to the contact point of the resultant *fracture vector* $\{\sigma_{fr}\} = \{\sigma\}^r + f_S^L \{\sigma\}^L$:

$$\sigma_1 = \sigma_1^r + f_S^L \sigma_1^L , \qquad \text{(Eq. 104)}$$

$$\sigma_2 = \sigma_2^r + f_S^L \sigma_2^L , \qquad \text{(Eq. 105)}$$

$$\tau_{21} = \tau_{21}^r + f_S^L \tau_{21}^L . \qquad \text{(Eq. 106)}$$

In this way we obtain equations whose only unknown is the desired stretch factor f_S^L of the load-determined stresses. After necessary reformulation, which in particular includes the removal of the root expressions, we obtain for each of the modes A, B, C a quadratic equation for f_S^L to which can be applied the known solution formula:

$$f_S^L = \frac{1}{2q}\left(\sqrt{l^2 - 4qc} - l\right). \qquad \text{(Eq. 107)}$$

Quantities q and l are the coefficients of the quadratic terms or of the linear terms of f_S^L, and in c are grouped the corresponding (constant) terms which are independent of f_S^L. Table 2 shows quantities q, l and c in each case for IFF modes A, B and C.

4.3.2.3 Procedure for calculating f_S^L and evaluation of results

With this task the starting situation is generally the following. The component design process came up with a laminate whereby specification of fiber directions and fiber quantities has been brought to a stage whereby the next step is to try to improve the laminate structure with regard to homogenizing the "safety" to IFF in the individual laminae. The load-determined stresses which need to be taken into account here in most cases relate to the maximum loading to be expected when the component is in service. With this maximum loading an addition safety margin with respect to IFF is to be provided and generally a larger safety margin with respect to FF. When considerable residual stresses are present, it is not possible on the basis of calculated stress exposure values f_{E_1} to come to any useful conclusions regarding these safety margins for the individual lamina, instead this is entirely possible on the basis of the stretch factors f_S^L of the load-determined stresses.

Before commencing with determining f_S^L with regard to IFF, it appears to be advisable to clarify whether an IFF can even occur in the lamina under consideration or whether a FF will appear first before an IFF.

From the simple FF condition given in (Eq. 14) it follows that it is solely stress σ_1 which is responsible for the FF. At FF:

$$f_{E,FF} = \frac{\sigma_1}{(\sigma_1)_{fr}} = \frac{\sigma_1^r + f_{S(FF)}^L \cdot \sigma_1^L}{(\sigma_1)_{fr}} = 1. \qquad \text{(Eq. 108)}$$

where $(\sigma_1)_{fr} = R_{\parallel}^t$ for $\sigma_1 > 0$
and $(\sigma_1)_{fr} = -R_{\parallel}^c$ for $\sigma_1 < 0$.

If σ_1^r and σ_1^L have the same sign, $\sigma_1 = \sigma_1^r + f_{S(FF)}^L \cdot \sigma_1^L$ can also only have the same sign. If σ_1^r and σ_1^L have different signs, it can be assumed that σ_1^r alone will not yet cause a FF. This is because a FF due solely to σ_1^r must not even be permitted and will have already been ruled out by the designer's pre-dimensioning. Should a stress $f_{S(FF)}^L \cdot \sigma_1^L$ with a different sign appear in addition to σ_1^r, then as it increases it will initially work in opposition to the residual stress σ_1^r. Not until after the "disappearance" of σ_1^r will the resulting stress σ_1 increase with the same sign as σ_1^L until the FF limit at $\sigma_1 = (\sigma_1)_{fr}$ is reached and FF finally occurs. Consequently the decision regarding the quantity $(\sigma_1)_{fr}$ to be used in equation (Eq. 108) can be made as follows:

$$\begin{aligned} (\sigma_1)_{fr} &= R_{\parallel}^t \text{ for } \sigma_1^L > 0 \\ (\sigma_1)_{fr} &= -R_{\parallel}^c \text{ for } \sigma_1^L < 0. \end{aligned} \qquad \text{(Eq. 109)}$$

The stretch factor $f_{s(FF)}^L$ of the load-determined stresses *with regard to FF* is thus obtained from equations (Eq. 108) and (Eq. 109):

$$f_{S(FF)}^L = \frac{(\sigma_1)_{fr} - \sigma_1^r}{\sigma_1^L}. \qquad \text{(Eq. 110)}$$

The three components σ_1^L, σ_2^L, τ_{21}^L of the load-determined stress vector are all increased by the same stretch factor. Once the FF limit is reached, the resulting stresses will therefore have these magnitudes:

$$\sigma_1 = \sigma_1^r + f_{S(FF)}^L \sigma_1^L, \qquad \text{(Eq. 111)}$$

$$\sigma_2 = \sigma_2^r + f_{S(FF)}^L \sigma_2^L, \qquad \text{(Eq. 112)}$$

$$\tau_{21} = \tau_{21}^r + f_{S(FF)}^L \tau_{21}^L. \qquad \text{(Eq. 113)}$$

With the aid of these stresses a check is made as to whether the contact point of the resulting stress vector (at $\sigma_1 = R_{\parallel}^t$ or $\sigma_1 = -R_{\parallel}^c$) lies inside or outside the contour line of the fracture body end face as shown in Fig. 50c. This contour line is the intersection line of the conical circumferential surface for IFF and the FF plane. It is a reduced (σ_2, τ_{21}) fracture curve for IFF which is geometrically similar reduced by the minimum weakening factor $\eta_{w1} = M$ valid at the FF limit.

With stresses σ_2 and τ_{21} calculated at the FF limit on the basis of equations (Eq. 112) and (Eq. 113) and $\eta_{w1} = M$, the IFF stress exposure with σ_1 influence can be calculated from:

$$f_{E_1} = \frac{f_{E_0}}{\eta_{W1}} = \frac{f_{E_0}}{M}, \qquad \text{(Eq. 114)}$$

doing so with the equation for mode A, B or C which is valid for the stress state under consideration (see equations (Eq. 71), (Eq. 73) and (Eq. 92)).

If the result $f_{E_1} = 1$ were obtained from equation (Eq. 114), the contact point would be located exactly on the IFF curve and simultaneously in the plane for FF. In theory this would therefore mean that both IFF and FF would occur. If $f_{E_1} < 1$ is the result, the vector passes through the FF plane without having previously touched an IFF circumferential surface – in other words, a f_S^L value cannot be calculated for IFF.

Should the result be $f_{E_1} > 1$, this means that the vector must pierce the IFF circumferential surface before it reaches the FF limit. In this case f_S^L is calculated for IFF using the procedure shown in the flow chart in Fig. 52.

Instead of the sequence shown in this flow chart, the computational operations and inquiries could also be carried out in a different order. Whatever the case, it ends up to a procedure whereby arbitrary assumptions are made first (for example, about which IFF mode is to be expected, A, B or C and $(\sigma_1)_{fr} = R^t_{\parallel}$ or $(\sigma_1)_{fr} = -R^c_{\parallel}$ as well as an assumption as to whether a σ_1 influence is already becoming effective or not. A decision is then made on the basis of the computational results as to whether the assumptions made were justified or not. If this does not appear to be the case, f_S^L is recalculated using modified assumptions. The procedure in the flow chart (Fig. 52) is that in each case the investigation is carried out for a selected mode A, B or C until the result is accepted or rejected. The final test is to check to see whether the calculated contact point actually falls within the range of validity of the equations associated with the mode under consideration.

If the result for a mode passes in addition to all proceeding tests also this final one, this means that the result is the correct and only possible result, because only one contact point exists.

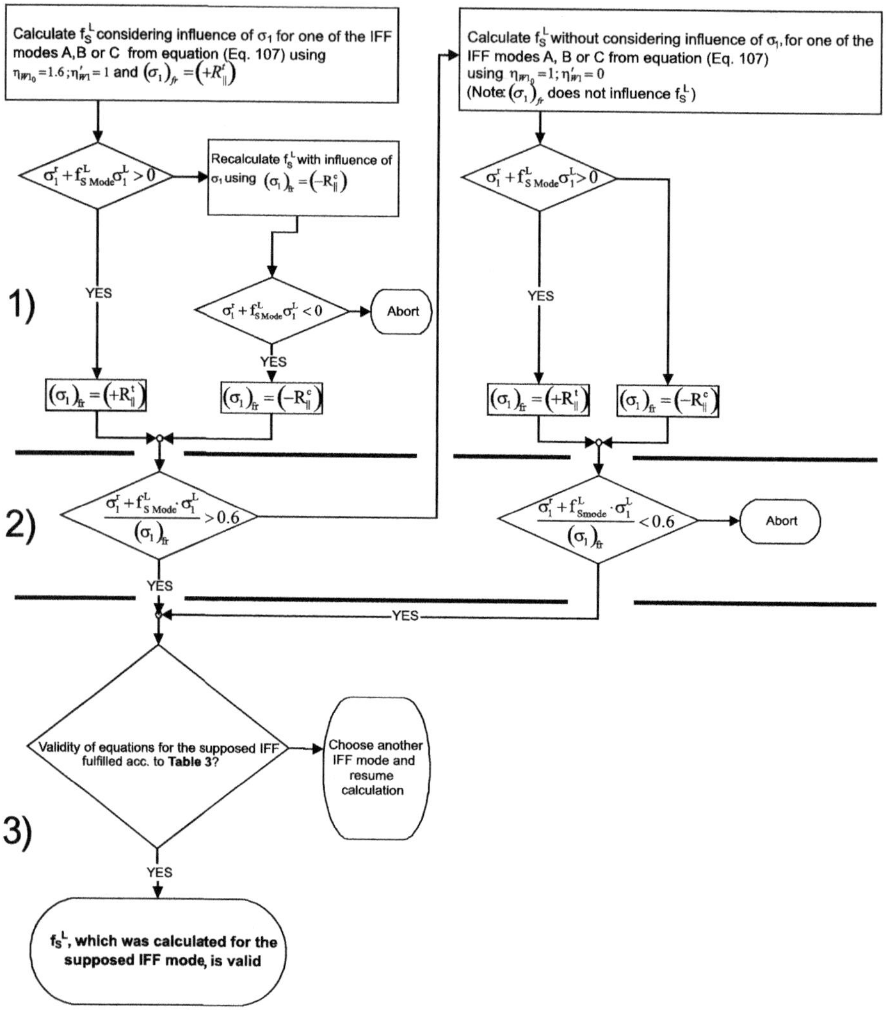

Fig. 52. Checks which have to be carried out during calculation of the stretch factor f_S^L for an IFF mode A, B or C in order to decide whether the weakening influence of σ_1 needs to be considered for the IFF modes, as well as the final check, whether the contact point to the fracture body surface falls within the range of validity of the equations for the considered IFF-mode A, B or C.

Table 2. Terms of equation (Eq. 107) for the calculation of f_S^L

IFF mode	
A	$q_A = \frac{\left(\tau_{21}^L\right)^2}{R_{\perp\parallel}{}^2} - \frac{2p_{\perp\parallel}^t \eta_{w1}^{'} \sigma_1^L \sigma_2^L}{R_{\perp\parallel}\left(\sigma_1\right)_{fr}} + \left(1 - \frac{2p_{\perp\parallel}^t R_\perp^t}{R_\perp^t}\right) \cdot \frac{\left(\sigma_2^L\right)^2}{\left(R_{\perp\parallel}\right)^2} - \frac{\left(\eta_{w1}^{'}\right)^2 \left(\sigma_1^L\right)^2}{\left(\sigma_1\right)_{fr}{}^2}$ $l_A = -\frac{2p_{\perp\parallel}^t \eta_{w1}^{'} \sigma_1^L \sigma_2^r}{R_{\perp\parallel}\left(\sigma_1\right)_{fr}} + \frac{2p_{\perp\parallel}^t \sigma_2^L}{R_{\perp\parallel}} \cdot \left(\eta_{w1_0} - \eta_{w1}^{'} \cdot \frac{\sigma_1^r}{\left(\sigma_1\right)_{fr}}\right) + \frac{2\sigma_2^r \sigma_2^L}{\left(R_\perp^t\right)^2} \cdot \left(1 - \frac{2p_{\perp\parallel}^t R_\perp^t}{R_{\perp\parallel}}\right) +$ $+ \frac{2\eta_{w1}^{'} \sigma_1^L}{\left(\sigma_1\right)_{fr}} \cdot \left(\eta_{w1_0} - \eta_{w1}^{'} \cdot \frac{\sigma_1^r}{\left(\sigma_1\right)_{fr}}\right) + \frac{2\tau_{21}^r \tau_{21}^L}{R_{\perp\parallel}^2}$ $c_A = \left(1 - \frac{2p_{\perp\parallel}^t R_\perp^t}{R_{\perp\parallel}}\right) \cdot \frac{\left(\sigma_2^r\right)^2}{\left(R_\perp^t\right)^2} - \left(\eta_{w1_0} - \eta_{w1}^{'} \cdot \frac{\sigma_1^r}{\left(\sigma_1\right)_{fr}}\right)^2 + \frac{\left(\tau_{21}^r\right)^2}{R_{\perp\parallel}^2} + \frac{2p_{\perp\parallel}^t \sigma_2^r}{R_{\perp\parallel}} \cdot \left(\eta_{w1_0} - \eta_{w1}^{'} \cdot \frac{\sigma_1^r}{\left(\sigma_1\right)_{fr}}\right)$
B	$q_B = \frac{\left(\tau_{21}^L\right)^2}{R_{\perp\parallel}{}^2} - \frac{2p_{\perp\parallel}^c \eta_{w1}^{'} \sigma_1^L \sigma_2^L}{R_{\perp\parallel}\left(\sigma_1\right)_{fr}} - \frac{\left(\eta_{w1}^{'}\right)^2 \left(\sigma_1^L\right)^2}{\left(\sigma_1\right)_{fr}{}^2}$ $l_B = \frac{2\tau_{21}^r \tau_{21}^L}{R_{\perp\parallel}^2} - \frac{2p_{\perp\parallel}^c \eta_{w1}^{'} \sigma_1^L \sigma_2^r}{R_{\perp\parallel}\left(\sigma_1\right)_{fr}} + \frac{2p_{\perp\parallel}^c \sigma_2^L}{R_{\perp\parallel}} \cdot \left(\eta_{w1_0} - \eta_{w1}^{'} \cdot \frac{\sigma_1^r}{\left(\sigma_1\right)_{fr}}\right) +$ $+ \frac{2\eta_{w1}^{'} \sigma_1^L}{\left(\sigma_1\right)_{fr}} \cdot \left(\eta_{w1_0} - \eta_{w1}^{'} \cdot \frac{\sigma_1^r}{\left(\sigma_1\right)_{fr}}\right)$ $c_B = \frac{\left(\tau_{21}^r\right)^2}{R_{\perp\parallel}^2} - \left(\eta_{w1_0} - \eta_{w1}^{'} \cdot \frac{\sigma_1^r}{\left(\sigma_1\right)_{fr}}\right)^2 + \frac{2p_{\perp\parallel}^c \sigma_2^r}{R_{\perp\parallel}} \cdot \left(\eta_{w1_0} - \eta_{w1}^{'} \cdot \frac{\sigma_1^r}{\left(\sigma_1\right)_{fr}}\right)$
C	$q_C = \frac{1}{4} \cdot \frac{\left(\tau_{21}^L\right)^2}{\left(R_{\perp\parallel} + p_{\perp\parallel}^c R_{\perp\perp}^A\right)^2} + \frac{\left(\sigma_2^L\right)^2}{\left(R_\perp^c\right)^2} - \frac{\eta_{w1}^{'} \sigma_1^L \sigma_2^L}{R_\perp^c \left(\sigma_1\right)_{fr}}$ $l_C = \frac{\tau_{21}^r \tau_{21}^L}{2\left(R_{\perp\parallel} + p_{\perp\parallel}^c R_{\perp\perp}^A\right)^2} - \frac{\eta_{w1}^{'} \sigma_1^r \sigma_2^L}{R_\perp^c \cdot \left(\sigma_1\right)_{fr}} - \frac{\eta_{w1}^{'} \sigma_1^L \sigma_2^r}{R_\perp^c \cdot \left(\sigma_1\right)_{fr}} + \frac{\eta_{w1_0} \sigma_2^L}{R_\perp^c} + \frac{2\sigma_2^L \sigma_2^r}{\left(R_\perp^c\right)^2}$ $c_C = \frac{1}{4} \cdot \frac{\left(\tau_{21}^r\right)^2}{\left(R_{\perp\parallel} + p_{\perp\parallel}^c R_{\perp\perp}^A\right)^2} + \frac{\eta_{w1_0} \sigma_2^r}{R_\perp^c} + \frac{\left(\sigma_2^r\right)^2}{\left(R_\perp^c\right)^2} - \frac{\eta_{w1}^{'} \sigma_2^r \sigma_1^r}{R_\perp^c \left(\sigma_1\right)_{fr}}$ **Note:** Because of the parameter coupling $p_{\perp\perp}^c R_{\perp\parallel} = p_{\perp\parallel}^c \cdot R_{\perp\perp}^A$ according to (Eq. 93) $R_{\perp\perp}^A$ needs to be calculated using (Eq. 94): $R_{\perp\perp}^A = \frac{R_{\perp\parallel}}{2p_{\perp\parallel}^c}\left(\sqrt{1 + 2p_{\perp\parallel}^c \frac{R_\perp^C}{R_{\perp\parallel}}} - 1\right)$
$\left(\sigma_1\right)_{fr}$	$\left(\sigma_1\right)_{fr} = R_\parallel^t$ for $\sigma_1 > 0$ $\left(\sigma_1\right)_{fr} = -R_\parallel^c$ for $\sigma_1 < 0$
	The relations given in Table 2 are also suitable for the calculation of f_S^L without influence of σ_1 by setting $\eta_{w1_0} = 1.0$ and $\eta'_{w1} = 0$

Table 3. Range of validity of the equations for IFF modes A, B, C

IFF mode	Range of validity
A	$\sigma_2^r + f_{S\,A}^L \cdot \sigma_2^L \geq 0$
B	$\sigma_2^r + f_{S\,B}^L \cdot \sigma_2^L < 0 \; and \; 0 \leq \left\lvert \dfrac{\sigma_2^r + f_{S\,B}^L \cdot \sigma_2^L}{\tau_{21}^r + f_{S\,B}^L \cdot \tau_{21}^L} \right\rvert \leq \dfrac{R_{\perp\perp}^A}{\lvert \tau_{21_c} \rvert}$
C	$\sigma_2^r + f_{S\,C}^L \cdot \sigma_2^L < 0 \; and \; 0 \leq \left\lvert \dfrac{\tau_{21}^r + f_{S\,C}^L \cdot \tau_{21}^L}{\sigma_2^r + f_{S\,C}^L \cdot \sigma_2^L} \right\rvert \leq \dfrac{\lvert \tau_{21_c} \rvert}{R_{\perp\perp}^A}$
	with $\tau_{21_c} = R_{\perp\parallel} \cdot \sqrt{1 + 2p_{\perp\perp}^c}$ ($p_{\perp\perp}^c$ according to *(Eq. 93)* and $R_{\perp\perp}^A$ according to *(Eq. 94)*)

With the coordinates σ_1, σ_2, τ_{21} of this contact point on the fracture body surface, the value $f_{E_1} = 1$ for the IFF stress exposure f_{E_1} must result from equation (Eq. 99) when applying the relevant equation (Eq. 101), (Eq. 102) or (Eq. 103). This can be used for checking purposes.

If a value is calculated for f_S^L which is <1, this indicates that the IFF limit will have already been exceeded with the existing load stresses $\sigma_1^L, \sigma_2^L, \tau_{21}^L$ used in the calculation!

4.4 Visualization of fracture bodies

The visualization of fracture conditions for FRP has been treated by Kopp [Kopp and Michali 1999]. This visualization can very much help to see and comprehend the characteristics of different fracture conditions. For the principal understanding it is important to recall the remarks on the co-ordinate systems in the chapter 3.

The visualization of fracture conditions for isotropic material is not difficult at all. In this case the 6 stress components σ_1, σ_2, σ_3, τ_{23}, τ_{31}, τ_{21} are substituted by 3 principal normal stresses σ_{IP}, σ_{IIP}, σ_{IIIP}. For fibrous transversely isotropic materials such a change of the coordinate systems is not possible, as a rotation is only allowed in the so called transversely isotropic (x_2, x_3)-plane. The x_1-axis must remain parallel to the fiber direction, because only for these 'natural directions' the basic strength parameters of the UD-layer are known.

The following investigations will only consider stresses having a direct influence on IFF. The fiber parallel stress σ_1 is disregarded, since IFF occurs in a fiber parallel plane [2, 3] and with that the fracture plane is parallel to σ_1. The corresponding situation for an isotropic material is that $\sigma_1 = \sigma_{IP}$ is neither the minimal nor the maximal principal normal stress. According to *Mohr* only the extreme values of the principal stresses σ_{IIP} and σ_{IIIP} do have an influence on the fracture process; for this reason fracture occurs parallel to the direction of σ_{IP} similar to the IFF in an UD-lamina.

For human imagination it is necessary to restrict fracture bodies to three dimensions. Therefore it seems reasonable in a first step to transform the (σ_2, σ_3, τ_{23}, τ_{31}, τ_{21})-stress state in consideration of the transverse isotropy into a (σ_{II}, σ_{III}, 0, τ_{III1}, τ_{II1})-stress state (Fig. 20). By no means σ_{II} and σ_{III} are 'real' principal stresses σ_{IIP} and σ_{IIIP}, because also the shear stresses τ_{III1}, τ_{II1} occur on their action planes (Fig. 20). Using the following transformation rule one is enabled to substitute any stress component appearing in a fracture condition in the way wanted:

$$\begin{Bmatrix} \sigma_2 \\ \sigma_3 \\ \tau_{23} \\ \tau_{31} \\ \tau_{21} \end{Bmatrix} = \begin{bmatrix} \cos^2\varphi & \sin^2\varphi & -2\cdot\sin\varphi\cdot\cos\varphi & 0 & 0 \\ \sin^2\varphi & \cos^2\varphi & 2\cdot\sin\varphi\cdot\cos\varphi & 0 & 0 \\ \sin\varphi\cdot\cos\varphi & -\sin\varphi\cdot\cos\varphi & \cos^2\varphi-\sin^2\varphi & 0 & 0 \\ 0 & 0 & 0 & \cos\varphi & \sin\varphi \\ 0 & 0 & 0 & -\sin\varphi & \cos\varphi \end{bmatrix} \cdot \begin{Bmatrix} \sigma_{II} \\ \sigma_{III} \\ 0 \\ \tau_{III1} \\ \tau_{II1} \end{Bmatrix} \qquad \text{(Eq. 115)}$$

$$\textit{with} \quad \varphi = \frac{1}{2}\arctan\frac{2\cdot\tau_{23}}{\sigma_2-\sigma_3}$$

The next step is to replace the shear stresses τ_{III1} and τ_{II1} by the resulting longitudinal shear stress $\tau_{\omega 1}$ as shown in (Fig. 53). The Figure clarifies that τ_{1III} and τ_{1II} are nothing else but the components of a longitudinal shear stress $\tau_{1\omega}$ acting on a plane perpendicular to the x_1-direction [Puck 1997]:

$$\tau_{1\omega} = \sqrt{\tau_{1III}^2 + \tau_{1II}^2} = \sqrt{\tau_{III1}^2 + \tau_{II1}^2} = \tau_{\omega 1} \qquad \text{(Eq. 116)}$$

The direction of $\tau_{1\omega}$ is defined by the angle δ:

$$\delta = \arctan\frac{\tau_{1III}}{\tau_{1II}} = \arctan\frac{\tau_{III1}}{\tau_{II1}} \qquad \text{(Eq. 117)}$$

δ is called the difference angle or deviation angle because its value is a measure for the deviation of the action planes of $\tau_{\omega 1}$ and σ_{II}. Stress states

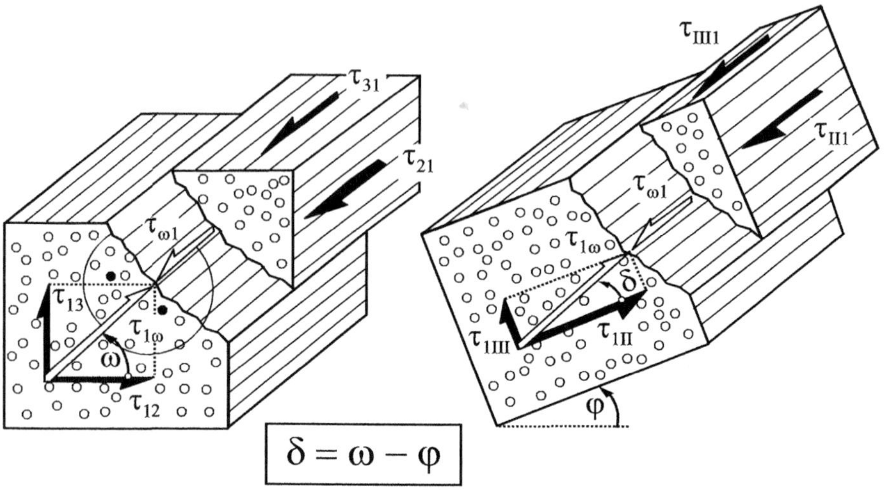

Fig. 53. Resulting shear stress $\tau_{\omega 1}$ and deviation angle δ

without transversal stresses σ_{II} and σ_{III} will lead to an IFF in the fiber parallel plane the resultant shear stress $\tau_{\omega 1}$ is acting in. In this special case the fracture angle is $\theta_{fp} = \omega = \arctan \tau_{31}/\tau_{21}$ and fracture takes place when $\tau_{\omega 1}$ has reached the longitudinal shear strength $R_{\perp\parallel}$. One can imagine that in general the deviation angle δ between the action planes of $\tau_{\omega 1}$ and σ_{II} is of main importance for the fracture process. It is obvious that in dependence of δ the interaction between the transverse extreme stresses σ_{II}, σ_{III} and the resulting shear stress $\tau_{\omega 1}$ is more or less distinct. As an example one can regard a ($\sigma_{II} > 0$, 0, $\tau_{\omega 1}$, δ)-stress state with $\delta = 0°$ on the one hand and $\delta = 90°$ on the other hand. In contrast to the complete interaction of σ_{II} and $\tau_{\omega 1}$ for $\delta = 0°$ there is no interaction at all for $\delta = 90°$ and fracture does not take place as long as either σ_{II} reaches the transverse tension strength $R_{\perp}^{t}$ or $\tau_{\omega 1}$ reaches $R_{\perp\parallel}$. It is inevitable to consider δ as a parameter for the visualization of fracture conditions. From an academic point of view this leads to an infinite number of fracture bodies in the (σ_{II}, σ_{III}, $\tau_{\omega 1}$)-stress space, as any small change of δ might modify the fracture body slightly. For the comparison of different fracture conditions it is absolutely sufficient to make a $\Delta\delta = 22{,}5°$ grading.

All fracture conditions visualized in this chapter were calculated with the same set of basic strength values published in [Cuntze et al. 1997]. If a criterion is complemented by additional material coefficients the suggestions of the accompanying developer were fulfilled.

4.4.1.1 Tsai, Wu-criterion

A classical representative of a global strength criterion is the well-known criterion of *Tsai* and *Wu* (Tsai and Wu 1971). It is a pure interpolation polynomial deduced from the basic strength values of the UD-layer, not regarding the fact that some strength values are dominated by fiber strength and others by matrix or interface strength.

The result is a smooth fracture body without any edges. Most certainly the fracture body does not change its shape if the influence of σ_I is neglected. According to expectations the *Tsai, Wu*-criterion takes on the shape of an ellipsoid in the (σ_{II}, σ_{III}, $\tau_{\omega I}$)-stress space, too (Fig. 54). There is no dependence on the deviation angle δ. As σ_{II} and σ_{III} are treated absolutely equally in the invariant formulation of the criterion the fracture body is moreover symmetrical with regard to the ($\sigma_{II} = \sigma_{III}$)-plane. Therefore stress states having a transverse normal stress and a longitudinal shear stress acting on the same plane are treated equally to those having their action planes perpendicular to each other. Another remarkable aspect of this fracture body is that it is closed for biaxial compressive stresses $\sigma_{II} \approx \sigma_{III}$. From a macro mechanical point of view – which is the basis of all strength criteria investigated here – the shear fracture due to transverse compression is not possible for $\sigma_{II} = \sigma_{III}$ as the maximum transverse shear

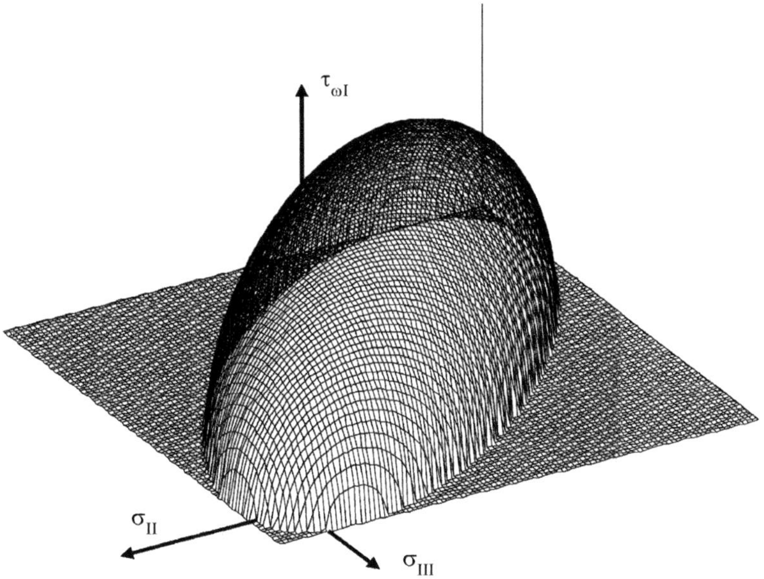

Fig. 54. Fracture body of the Tsai,Wu-criterion

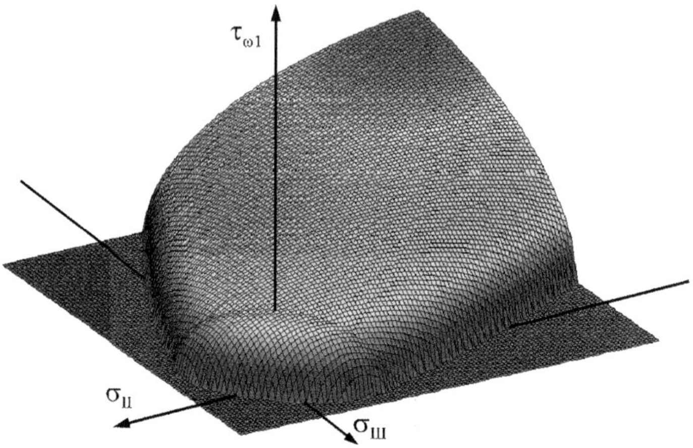

Fig. 55. Fracture body of the Hashin-criterion

stress $\tau_{\perp\perp max} = ½\cdot(\sigma_{II} - \sigma_{III})$ always remains zero. However – because of micromechanical inhomogeneity – a 'damage limit' for biaxial transverse compressive stresses must be expected ([Kopp et al. 1997]). After exceeding this 'damage limit' the micro-cracks in the matrix will reduce the original mechanical properties noticeably but a real material separation does not happen.

4.4.1.2 Hashin-criterion

Usually the term Hashin-criterion for IFF is related to two invariant fracture conditions published in (Hashin 1980), although in the same publication an action plane related approach is discussed too, which gave Puck the impulse to develop his action plane related IFF-criterion (Puck 1992). As Hashin was aware of the importance of the sign of the normal stress σ_n on the fracture plane, his invariant formulations differentiate between the cases $\sigma_n \geq 0$ and $\sigma_n < 0$. Unfortunately the fracture plane is not known a priori and cannot be found by using invariant formulations. Therefore Hashin had to define a more practical condition to delimit his fracture conditions from each other. He fixed the boundary of the sections for $\sigma_n \geq 0$ and for $\sigma_n < 0$ in the plane $\sigma_{III} = -\sigma_{II}$. However he simultaneously pointed out the severe physical contradictions, which result from this rather arbitrary boundary.

The corresponding fracture body is composed of a paraboloid with elliptical cross sections for $\sigma_n < 0$ and a crosswise oriented ellipsoid for $\sigma_n \geq 0$. Both are symmetric to the $(\sigma_{II} = \sigma_{III})$-plane and intersect each other in the

($\sigma_{III} = -\sigma_{II}$)-plane in an elliptical borderline. The fracture curve for $\tau_{\omega 1} = 0$ and $\sigma_n \geq 0$ contradicts to the theory of Paul (Paul 1961), which assumes no interactions of the principal stresses in this area, but Hashin already takes into account that a real fracture due to biaxial transverse compression is not possible. Accordingly the fracture body is not closed for $\sigma_{II} \approx \sigma_{III}$. Again it has to be emphasized that the Hashin fracture body is independent of the deviation angle δ.

4.4.1.3 Puck criterion

Figure 56 displays the fracture bodies of Puck's IFF-criterion in the (σ_{II}, σ_{III}, $\tau_{\omega 1}$)-stress space in dependence on δ. Of course all fracture bodies show the same base in the (σ_{II}, σ_{III})-plane, which has – except of the compression/compression domain – the shape of Paul's fracture envelope for intrinsically brittle isotropic material (Paul 1961). Comparable to Hashin's approach in the third quadrant fracture due to biaxial transverse compression is not possible.

Particularly interesting is the shape of the fracture bodies for $\delta = 0°$ and $\delta = 90°$. Here the components of the fracture surface, which belong to the different fracture modes marked in the figure, touch each other in sharp edges. (A fracture mode is defined by the combination of stressings acting on the fracture plane (Puck 1997)). It is remarkable that *Puck's* fracture

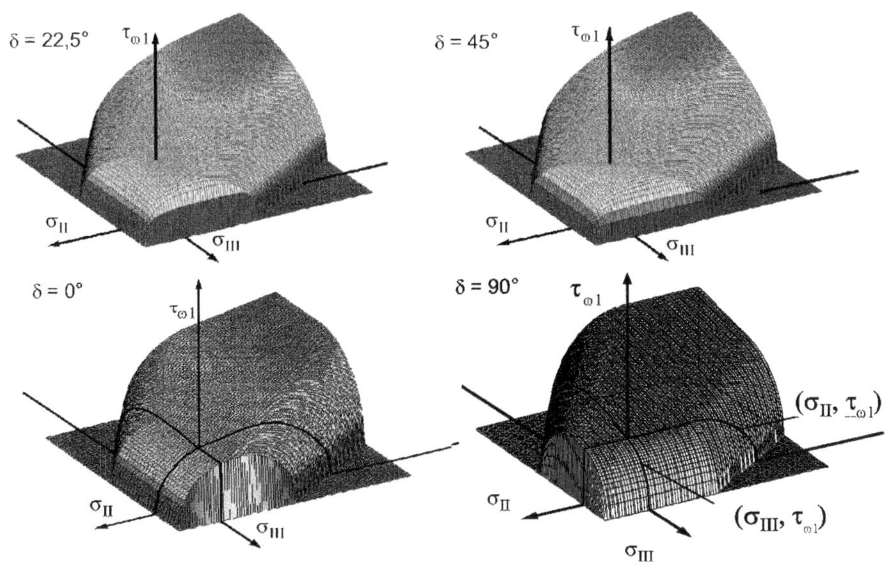

Fig. 56. Fracture bodies of the action plane related criterion by Puck

conditions, which describe a smooth ellipsoid combined with a smooth paraboloid in the (σ_n, τ_{nt}, τ_{n1})-stress space (compare Fig. 45), have the ability to describe so many fracture surface portions in the (σ_{II}, σ_{III}, $\tau_{\omega 1}$)-stress space. The reason for this is given by the search of the plane with the highest stress exposure factor. Thereby the criterion corresponds in certain sections to a maximum stress criterion, whereas other stress combinations are assessed in the form of a mixed-mode criterion. These effects can be clarified excellently by means of the fracture envelopes for $\sigma_{II} = 0$ and $\sigma_{III} = 0$ which have been drawn onto the fracture body for $\delta = 90°$. Whereas the (σ_{III}, $\tau_{\omega 1}$)-fracture envelope corresponds in its complete extent of validity to a mixed-mode criterion, by far the most of the (σ_{II}, $\tau_{\omega 1}$)-fracture envelope corresponds to a maximum normal stress criterion or a maximum longitudinal shear stress criterion. The explanation for this phenomenon is, that in case of a deviation angle $\delta = 90°$ the stresses $\tau_{\omega 1} = \tau_{III1}$ and σ_{III} are acting on the same fiber parallel plane and therefore promote fracture together. In contrast $\tau_{\omega 1} = \tau_{III1}$ is acting on a plane that is perpendicular to the one σ_{II} is acting on; according to *Mohr* an interaction is therefore not possible until an inclined fracture plane occurs and the transverse shear stress τ_{nt} due to high σ_{II}^c co-operates with the longitudinal shear stress τ_{n1} to produce fracture.

Passing over to deviation angles, which differ some degrees from 0° and 90°, the sections comparable to a maximum stress criterion vanish, except for the (σ_{II}, σ_{III})-base. In case of $\delta = 45°$ the fracture body has to be symmetrical with regard to the ($\sigma_{II} = \sigma_{III}$)-plane, as the action plane of $\tau_{\omega 1}$ is equally inclined towards the action planes of σ_{II} and σ_{III}.

The comparison of fracture bodies in the (σ_{II}, σ_{III}, $\tau_{\omega 1}$, δ)-stress space clearly demonstrates the shortcomings of criteria that do not account for the deviation angle δ between the action planes of the transverse principal stress σ_{II} and the resulting longitudinal shear stress $\tau_{\omega 1}$.

4.5 Summary of chapter

This chapter is the core part of the whole book. Here the Puck criteria are derived and presented in all detail. It is worth to read it from the first line on in order to fully understand the physical background of the criteria and the mathematical derivation.

Puck's criteria are based on the concept of intrinsically brittle materials. As an introduction Mohr's fracture hypotheses, the concept of principal stresses and the visualization in the Mohr circles are recalled and extended to orthotropic materials.

The fiber fracture criteria are presented prior to the IFF-criteria, because their explanation takes less space and the derivation of the criteria is simpler. Puck has developed FF-criteria which take all possible influences of shear and transverse normal stresses into account. However, for most cases the most simple maximum stress criterion is still valid for FF-analysis.

The main focus of Puck's work has always been on Inter Fiber Fracture. Puck's criteria offer not only a perfect calculation of the fracture limit and the stress exposure. In addition the fracture angle is calculated. With this additional information the differentiation between tolerable and non-tolerable IFF becomes possible.

The IFF-criteria are first explained in their general 3D-formulation. The mathematical derivation is made step by step and easy to follow. From the general formulation the equations for plane 2D-states of stress are derived. These 2D-formulations are of interest because they do not need the numerical search of the fracture plane.

Puck's criteria are based on knowledge of the idea that fracture is caused by the stresses on the fracture plane. Therefore in the original formulation of Puck's criteria no influence of fiber parallel stress on IFF is reflected. However, as a matter of fact high σ_1 stress leads to microdamage and reduces the strength of the material against IFF. Puck has developed a way to incorporate this into his criteria. In the same way he has incorporated probabilistic effects. These later developments are discussed under the headline "Extensions to the IFF-criteria". Another chapter is dedicated to the visualization of (IFF-) criteria. This visualization is a very good means to illustrate the characteristics of a creation and to compare different criteria. This comparison is done here for the classical Tsai/Wu-criterion, the Hashin-criterion and the Puck criteria.

5 Analysis of the gradual failure process

Fiber Fracture usually leads to total laminate failure whereas IFF can often be considered as tolerable damage (depending on the mode of fracture and the function of the designed part). Well known examples for parts with IFF as tolerated damage are torsion springs, pressure vessels and gas tanks. These tanks have a superior light weight potential compared to conventional steel vessels. However, this potential can just be exploited by tolerating IFF in the load carrying laminate. The macroscopic cracks develop spontaneously when the vessel is loaded the first time and run through the whole thickness of a layer being stopped at the fibers of a neighboring layer. For an optimal design of these high performance components it is essential to take the occurrence of IFF into account during the design process.

To do so, it must be analyzed first, what IFF means to a multilayer laminate. The damage caused by $IFF^{Mode\ A}$ (caused by transverse tensile stressing $\sigma_{\perp}^{t}$ and shear stressing $\tau_{\perp\parallel}$) leads primarily to a loss of in-plane stiffness of the cracked layer. However, as already explained in the chapter "Laminate Failure" a lamina damaged by IFF does not loose its transverse stiffness completely. At a distance from the crack, load is re-introduced into the lamina by shear stresses acting between the adjacent laminae (compare chapter "Laminate Failure") and from a macroscopic point of view the lamina behaves like an undamaged lamina with reduced moduli $E_{\perp}$ and $G_{\perp\parallel}$. If the load is further increased, new cracks can occur in the lamina with the result of a further degradation of $E_{\perp}$ and $G_{\perp\parallel}$. Both numerical models and experiments show that there is a maximum crack density. Once this "characteristic crack density" is reached, further increase in load leads to other forms of damage in the laminate, mainly delamination between the affected lamina and the neighboring laminae. This delamination starts from the tip ends of the IFF-cracks.

Damage in other layers of the laminate can of course occur long before the characteristic crack density has been reached. The degradation of the moduli $E_{\perp}$ and $G_{\perp\parallel}$ leads to a redistribution of stresses in the whole laminate. As a consequence further damage in other layers or even laminate failure might result without increasing the external load (Fig. 57). This damage leads to a redistribution of stresses in the laminate, too, and might

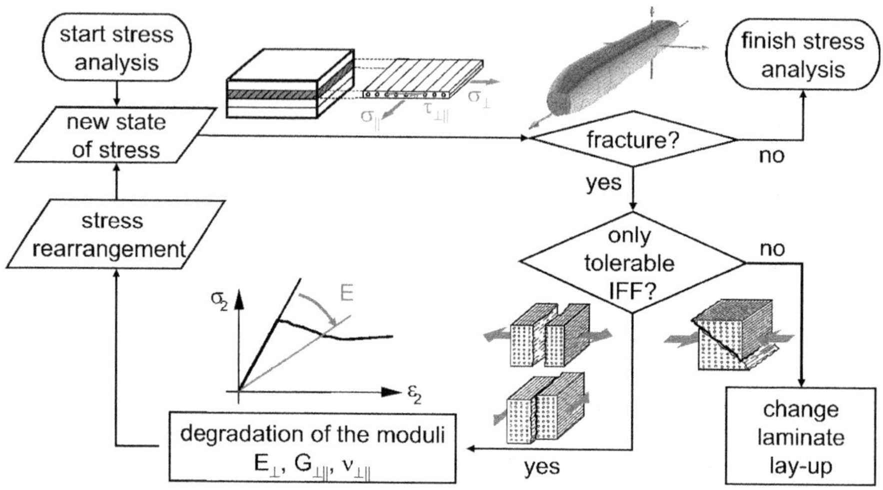

Fig. 57. Scheme of the gradual failure process

itself provoke further damage. This well known phenomenon is called gradual failure process. An analysis taking the gradual failure process into account is understood as “post failure analysis” even though the term might be regarded as not perfectly fitting as the laminate is not defined as “failed” just because of the IFF occurring.

Until now, the application of the described lamina by lamina fracture analysis of the laminate is limited to laminates made up of UD laminae and to intralaminar stresses – in other words, to two dimensional stress states (σ_1, σ_2, τ_{21}) in the laminae. Delamination cannot be described in this way. The restriction to 2-dimentional stress states (σ_1, σ_2, τ_{21}) applies to all the models and implementations described below.

Post failure analysis concentrates so far on monotonously increasing loads. However, the gradual failure process under these loads is relevant for dynamically loaded parts, too. Up to now, a true proof of the materials strength under cyclic loading is just possible by long-term fatigue testing [VDI 2006]. Meanwhile, there are numerous S-N-curves available in literature, but they are only valid for the laminate lay-up tested. Changes in the stacking sequence, layer thickness and fiber direction lead to totally different S-N-curves. A real improvement of the life time analysis of laminates is only possible on the basis of a profound understanding of composite materials and a clear differentiation between the different damage mechanisms. The analysis of the gradual failure process is very useful in this context.

5.1 Approaches for the modeling of the gradual failure process

In principle the gradual failure process as illustrated in Fig. 57 can be modeled by fracture mechanical-, damage mechanical- and phenomenological approaches. In the following a brief overview over the different approaches for modeling the gradual failure process is given before Puck's approach is explained in more detail.

Fracture mechanical approaches offer the advantage of potentially covering effects of stacking sequence and layer thickness [Nairn et al. 1993]. However, the modeling of laminates with IFF in laminae of different fiber direction is so far not possible with fracture mechanical means [Nairn 2000, Talreja 1994]. Thus the application of fracture mechanical approaches is generally limited to simple load cases and few laminate lay-ups. This statement is underlined by the results of the "World Wide Failure Exercise" [Soden et al. 1998; Hinton et al. 2002], where the most prominent experts in the field of composite failure analysis were asked to apply their own theories and models to a given set of test cases. In a second part of the contest the theoretical predictions were compared to experimental results supplied by the organizers of the Exercise. The participating fracture mechanical approaches could just deal with very few of the test cases [Mc Carney 1998]. Effectively, these were simple cross-ply laminates $[0/90]_S$ under uniaxial load [Mc Cartney 2002, Hinton et al. 2002].

Damage mechanical approaches generally consist of two parts: a damage representation and a damage growth law. The damage representation describes the correlation between a damage parameter ω (usually related to the crack density) and the degradation of the mechanical parameters. There is a huge variety of – mostly micro-mechanical – models [Chang and Chang 1987] which usually work fine. However, the second essential part of a damage mechanical model for the gradual failure process is the damage growth law. The state of the art in the field of damage growth laws is – in contrast to the field of damage representation – underdeveloped [Li et al. 1994]. This is due to both a lack of experimental data and of sophisticated models. One of the best damage mechanical models known to the author is that of *Li*, *Reid* und *Soden* [Li et al. 1998], because it implies a sophisticated and well working damage growth law.

Phenomenological approaches do not use a damage representation law which correlates the crack density with the degradation of mechanical parameters. Instead, these models are based on a stress/strain-analysis with the Classical Laminate Theory (CLT) or 3D-formulations, respectively, and a fracture criterion differentiating at least between Fiber Fracture and Inter Fiber Fracture. The degradation of mechanical properties must be known as a function of lamina-stresses and strains. The necessary data can generally be obtained by an adequate testing of specimens. Alternatively, some authors have chosen arbitrary degradation curves in the past, simply degrading parameters correlated to IFF to zero or a fixed value after IFF-initialization [Sun and Tao 1998, Liu and Tsai 1998, Gotsis et al. 1998, Wolf and Butalia 1998]. Such a ply-discount-method obviously represents a worst-case estimation of the influence IFF have on the laminate. A realistic analysis of the damage is not possible with this simple method. In the past several authors have worked on the improvement of this status quo [Nahas 1986, Soden et al. 1998]. The model of Puck [Puck 1969, Puck 1996, Puck and Schürmann 2002] which has been further developed by the author [Knops 2003] is considered a very sophisticated and realistic model.

Both the damage mechanical model of Li et al. [Li et al. 1998] and the phenomenological models are neglecting the influence of the stacking sequence and the lamina thickness on the fracture limit and the degradation. This might be justified by parametric studies of Li showing that the damage parameter ω as a function of the crack density is a fixed material parameter for a given UD-lamina which does not change considerably with lamina thickness and laminate lay-up [Thionnet and Rennard 1993]. Recent FEM-studies come to the same result [Lambrecht 2007]. However, these results are controversial and in clear contradiction to the results of Flaggs and Kural [Flaggs and Kural 1982]. Experimental results from Garrett and Bailey [Garret and Bailey 1977, Bailey et al. 1979] show, that an effect of the lamina thickness can only be measured if the 90° layers – of a $[0/90]_s$-laminate loaded in 0°-direction – are considerably thinner than the 0°-layers. All in all the controversial discussion shows that further experimental work is urgently needed in this field.

In contrast to this it is non controversial that layers with a free surface (bottom and top layer of a laminate) behave different than embedded layers. In the bottom or top layers the reduction of averaged stresses due to IFF cracks is stronger than in embedded layers. Therefore, on the one hand the same crack density leads to a stronger degradation of the stiffness of the layer [Nairn and Hu 1992]. On the other hand, the crack density in bottom and top layers is lower than in embedded layers at the same load [Li et al. 1998]. According to [Smith et al 1998] both effects completely compensate each other if the damage is considered as a function of the stress exposure f_E (measure for the risk of IFF-formation) and not of the crack density.

Both, highly developed damage mechanical models like that of Li et al. [Li et al. 1998] and phenomenological models like that of Puck are suitable to model the gradual failure process in arbitrary laminates. Any of these models requires experimental data for either calibration (phenomenological models like that of Puck) or verification (damage mechanical models like that of Li).

5.2 Puck's approach for the analysis of the gradual failure process

The basis of Puck's degradation model is the Inter Fiber Fracture analysis with an adequate model. The use of the Puck IFF-criteria is recommended, but it is not mandatory. In principle all criteria which determine the IFF-limit correctly can be used. The underlying stress analysis for the gradual failure analysis is usually conducted with the Classical Laminate Theory (CLT). However, attention has to be put on certain aspects. If the analyzed component is loaded by bending or torsion, the layers should be subdivided in very thin laminae in order to arrive at accurate laminae stresses. Besides, non-linearities of $E_{\perp}$ and $G_{\perp\parallel}$ as discussed in the Annex (chapter "Nonlinear stress analysis before IFF") should be taken into account. This means, that instead of using the initial engineering constants of Young's modulus $E_{\perp k}$ and the shear modulus $G_{\perp\parallel k}$ for a lamina k, the corresponding secant moduli (as a function of the stress σ_2 or τ_{21} respectively) are more suitably employed.

When determining the secant moduli $E_{\perp s}$ it is important to consider the Poisson effect. There is no explicit correlation between $E_{\perp s}$ and ε_2. However, there is such an explicit correlation between a "corrected" strain $\varepsilon_{2,1}{}^{comb}$ (eliminating Poisson strains) and $E_{\perp s}$. This is explained in detail in the Annex (chapter "Nonlinear stress analysis before IFF"). Meanwhile, deformation measurements on laminae under (σ_2, τ_{21}) stress combinations revealed that σ_2 has also an influence on the (τ_{21}, γ_{21}) curve and that on the other hand τ_{21} does influence the (σ_2, ε_2) curve [Puck 1996, Kopp et al. 1997, Kopp2000]. In a more recent paper a computational approach has been presented to take these effects into account by means of a simple physically based model (compare Annex, [Puck and Mannigel 2007]).

Furthermore, the residual stresses need to be considered correctly in the stress analysis. In most software tools the residual stresses are calculated separately to the load stresses and added to those as a constant value. However, in reality, the residual stresses reduce with increasing load stresses. This – at first glance not obvious effect – can be illustrated with the following thought experiment: A thin UD-laminae is – for instance by a shear frame – deformed to a shear strain $\gamma_{21}{}^{r}$. In a next step, other – thick and very stiff – layers are bonded on both sides of the deformed UD-lamina. Now a further shear deformation is added to the laminate (compare

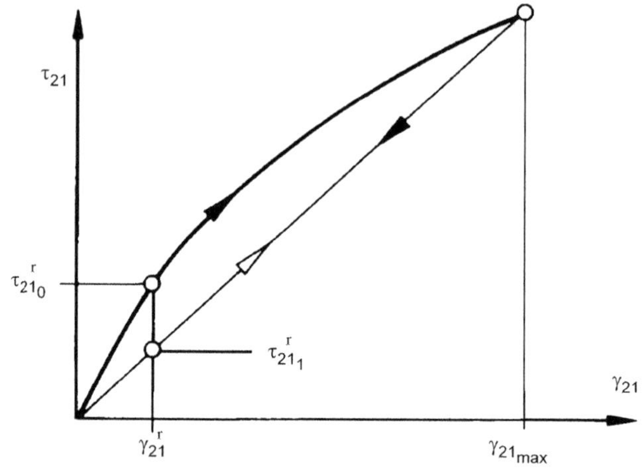

Fig. 58. Decreasing residual stress $\tau_{21}{}^r$ during a load hysteresis

Fig. 58). After in the layer under consideration a maximum shear deformation γ_{21max} in the non-linear part of the stress/strain-diagram has been reached, the laminate is unloaded completely. It is assumed, that only the thin, predeformed lamina has suffered micro–damage. In this case, the thin lamina arrives at the original (pre-)deformation $\gamma_{21}{}^r$. This time however at a lower shear stress $\tau_{21}{}^r{}_1$. Obviously, the micro-damage which occurred during the load cycle has reduced the original residual stress. In fact, the residual stress was already reduced to the lower value $\tau_{21}{}^r{}_1$ when the maximum deformation γ_{21max} was reached. As a consequence, in the course of the stress and strength analysis of a laminate the residual stresses in the laminae have to be recalculated for every load step using the (degraded) secant moduli $E_{\perp s}$ and $G_{\perp\| S}$.

Generally, if the described non-linear effects are not included in the stress analysis, lower IFF-limits are calculated.

For the degradation analysis it is necessary to distinguish whether there is transverse tensile stressing ($\sigma_\perp{}^t$, $IFF^{Mode\ A}$) or transverse compressive stressing ($\sigma_\perp{}^c$, $IFF^{Mode\ B}$ or $IFF^{Mode\ C}$) in the cracked lamina. Therefore, in the following chapters the degradation procedure is presented separately for cracks due to $IFF^{Mode\ A}$ and $IFF^{Mode\ B,\ C}$ respectively.

5.2.1 Degradation procedure for cracks due to $IFF^{Mode\ A}$

Cracks due to $IFF^{Mode\ A}$ tend to open because the fracture angle is $\theta_{fp} = 0°$ and the stress $\sigma_n = \sigma_2$ perpendicular to the fracture plane is a tensile stress.

That means that at the location of a crack the moduli $E_\perp$ and $G_{\perp\parallel}$ drop to zero. However, at a distance from the cracks – due to interlaminar shear – the cracked layer takes over load from the neighboring layers and stresses σ_2 and τ_{21} are developing again. In these regions between the cracks moduli $E_\perp > 0$ and $G_{\perp\parallel} > 0$ can be assumed. For a more macroscopic calculation, the uneven stress distribution in the cracked lamina is approximated by a constant stress which has the average value of the uneven stresses. This means that the effect of the cracks is "smeared" over the length dimension. In calculating practice, this is achieved by using gradually reduced secant moduli $E_\perp$ and $G_{\perp\parallel}$ after IFF onset. Originally, Puck believed that $E_\perp$ and $G_{\perp\parallel}$ should be degraded by the same factor η, where η is a function of the crack density. Experimental evidence as well as numerical studies (compare chapter "Experimental work") showed however, that the shear modulus $G_{\perp\parallel}$ is less affected by IFF-cracks than the Young's modulus $E_\perp$.

5.2.2 Degradation procedure for cracks due to $IFF^{Mode\ B}$ and $IFF^{Mode\ C}$

Developing a reasonable degradation process for IFFs of Mode B and C is much more difficult than that for $IFF^{Mode\ A}$. Based on the hypothesis that brittle IFF is influenced only by the stresses σ_n, τ_{nt}, τ_{n1} on the fracture plane, it is assumed that in Mode B ($\theta_{fp} = 0°$) a shear fracture occurs which is produced by $\tau_{n1} = \tau_{21}$. In Mode C ($\theta_{fp} \neq 0°$) the combined action of $\tau_{nt} = -\sigma_2 \sin\theta_{fp} \cos\theta_{fp}$ and $\tau_{n1} = \tau_{21} \cos\theta_{fp}$ causes a shear fracture on an inclined fracture plane. In both modes the simultaneously acting compressive stress $\sigma_n = \sigma_2$ (Mode B) or $\sigma_n = \sigma_2 \cos^2\theta_{fp}$ (Mode C) impedes the shear fracture.

In Mode B and C the cracks cannot open as in Mode A. Instead, the crack surfaces are pressed on each other. Therefore, it seemed to be reasonable to apply no η-degradation to $E_\perp$. In spite of the fact that cracks exist, the material is likely to behave like an undamaged (bulk) material as long as the compressive stress presses the crack surfaces on each other. Unlike $E_\perp$, the shear modulus $G_{\perp\parallel}$ undergoes some η-degradation because some sliding movement of the crack surfaces relative to each other is to be expected. This is, however, physically not completely correct, because before any relative sliding movement of the crack surfaces a certain friction has to be overcome. In general the degradation caused by a certain crack density will be less significant than for Mode A cracks.

Experimental practice has shown meanwhile that the real behavior of the material is even more complex. Mode B cracks tend to open at moder-

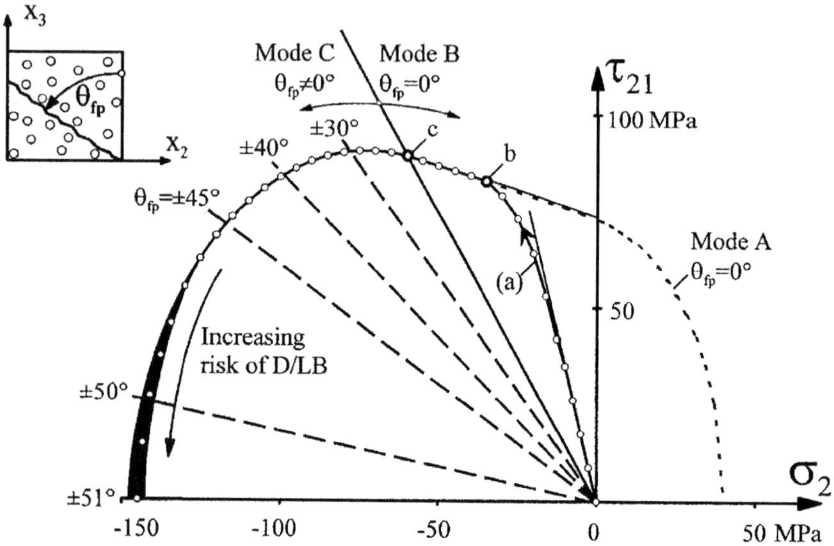

Fig. 59. Change from Mode B to Mode C with increasing load

ate σ_2^c-stress in spite of the load induced compressive stress on the crack surface. This is due to tensile curing stresses acting in the material perpendicular to the fibers. The resulting elastic behavior is documented in chapter "Experimental determination of degradation curves".

It is important to be aware of the fact, that, if a certain load path results in cracks of $IFF^{Mode\ B}$ in a layer, it is well possible that the fracture mode changes to Mode C if the load is increased further (compare Fig. 59). This is due to the different degradation of $G_{\perp\parallel}$ and $E_{\perp}$. As described earlier, Mode C cracks potentially lead to the so called "wedge effect" (compare Fig. 8). With high angles θ_{fp} between the fractured surface and the thickness direction, the fractured pieces act like wedges. If the fractured layer is a relative thick one placed in the middle of the laminate, the wedge effect may cause the laminate to "explode" causing severe delamination followed by local buckling of outer layers. This effect can only occur on inclined shear fracture planes with an angle θ_{fp} of at least ±30°. If such a high fracture angle is calculated, the risk of delamination and local buckling justifies regarding the laminate as failed, even though it might still carry the load and even take additional load.

5.2.3 Puck's method from 1969

According to Puck's original degradation method from 1969 [Puck 1969], the degradation of the moduli $E_\perp$, $G_{\perp\parallel}$ (and $\nu_{\perp\parallel}$) is governed by the stress exposure $f_{E,\,IFF}$. Puck's model describes the degradation of $E_\perp$, $G_{\perp\parallel}$ and $\nu_{\perp\parallel}$ according to (Eq. 118).

$$\begin{pmatrix} E_\perp \\ G_{\perp\parallel} \\ \nu_{\perp\parallel} \end{pmatrix} = \begin{pmatrix} \eta_E \cdot E_{\perp s} \\ \eta_G \cdot G_{\perp\parallel s} \\ \eta_\nu \cdot \nu_{\perp\parallel s} \end{pmatrix} \tag{Eq. 118}$$

Here $E_\perp$ and $G_{\perp\parallel}$ and $\nu_{\perp\parallel}$ are the degraded moduli, $E_{\perp s}$ and $G_{\perp\parallel s}$ the secant-moduli at IFF-initiation. The degradation factors η_E, η_G and η_ν are defined according to (Eq. 119)

$$\begin{aligned} \eta_E &= \frac{1-\eta_{rE}}{1+c_E\left(f_{E,IFF}-1\right)^{\xi_E}} + \eta_{r\,E} \\ \eta_G &= \frac{1-\eta_{rG}}{1+c_G\left(f_{E,IFF}-1\right)^{\xi_G}} + \eta_{r\,G} \\ \eta_\nu &= \frac{1-\eta_{r\nu}}{1+c_\nu\left(f_{E,IFF}-1\right)^{\xi_\nu}} + \eta_{r\nu} \end{aligned} \tag{Eq. 119}$$

where η_r, c and ξ are parameters which need to be calibrated by experimental data (Puck 1969). The parameter η_r has the physical meaning of a residual stiffness of the lamina once the characteristic damage stage is reached. The other two parameters are pure fitting parameters. Figure 60 shows the principle course of η as a function of the stress exposure.

Originally Puck proposed to degrade $E_\perp$, $G_{\perp\parallel}$ and $\nu_{\perp\parallel}$ with a common factor η. However, both modeling with fracture mechanical approaches and experimental data have shown that at a given crack density (and a given value of $f_{E,\,IFF}$) $E_\perp$ degrades to a larger extent than $G_{\perp\parallel}$ and that the major Poisson ratio $\nu_{\perp\parallel}$ does not need to be degraded at all [Knops 2003].

In principle other mathematical formulations than those of (Eq. 119) can be used, too, to define $\eta(f_{E,\,IFF})$ as long as these functions are adequate to define a curve as it is shown in Fig. 60 and as long as they can be fitted to experimental data. The degradation parameters η are equal 1 for $f_{E,\,IFF} \leq 1$ and smaller than 1 for $f_{E,\,IFF} > 1$.

The stress exposure factor $f_{E,\,IFF}$ is calculated using the secant-moduli $E_{\perp s}$, $G_{\perp\parallel s}$ and $\nu_{\perp\parallel s}$ at IFF-initiation. The stresses calculated from the lami-

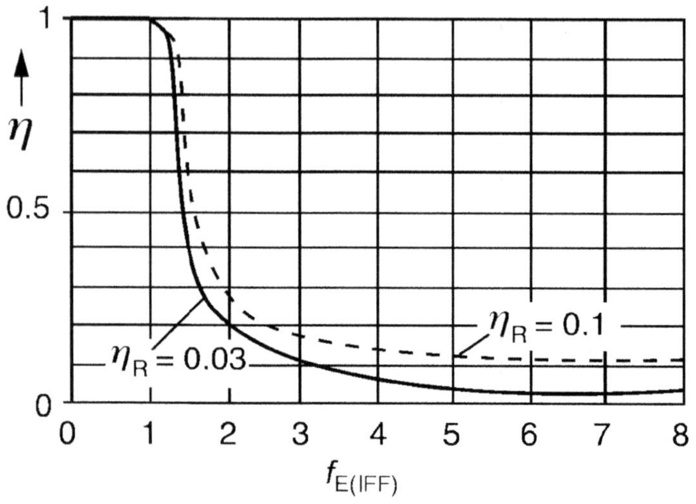

Fig. 60. Degradation factor η as a function of $f_{E,\,IFF}$

nate strains with these moduli are higher than the "real" stresses in the lamina. The use of these "fictive" stresses is a means for the definition of a straight forward degradation routine. As long as there is a known correlation between laminate load and crack density δ, the degradation η can both be expressed as η(δ) and $\eta(f_{E,\,IFF})$. The two degradation functions are absolutely equivalent (compare chapter "Experimental determination of degradation curves"). However, the correlation between load and crack density is only known in well documented experimental work. In most papers, only the correlation between crack density and stiffness degradation is given and the correlation between load and crack density is missing.

To make the degradation a function of the stress exposure is the key characteristic of Puck's post failure analysis. By this means a realistic degradation and consequently a realistic post failure analysis becomes possible without means of fracture or damage mechanics. However, the critical part is the correct definition of the degradation functions η_E, η_G and η_ν. This definition has to be done by experiments.

Extensive work to provide experimental data for the calibration of post failure models was carried out by the author [Knops 2003]. This is presented in detail in the chapter "Experimental work". Table 4 shows the Puck degradation parameters (compare (Eq. 119)) derived from the experimental results of Knops. The corresponding curves are shown in Fig. 61.

Table 4. Recommended parameters for the degradation function η [Knops Bögle 2006]

Material	GFRP		CFRP	
Parameter	$E_\perp$	$G_{\perp\parallel}$	$E_\perp$	$G_{\perp\parallel}$
Parameter valid for $\sigma_\perp \geq 0$				
c	5.3	0.70	5.3	0.95
ξ	1.3	1.5	1.3	1.17
η_r	0.03	0.25	0.03	0.67

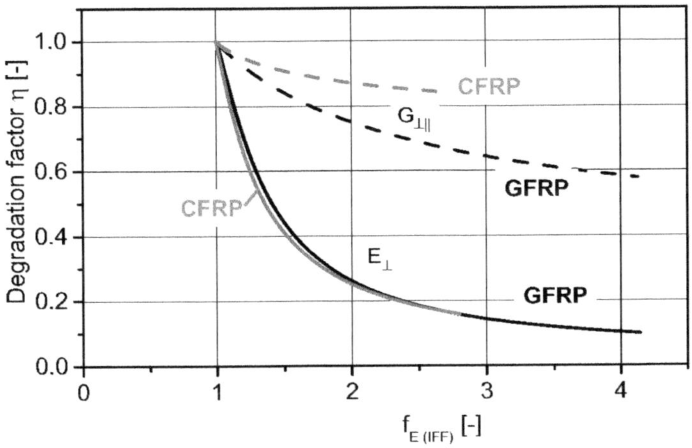

Fig. 61. Experimentally determined degradation curves as a function of $f_{E,IFF}$ [Knops Bögle 2006]

Regarding Fig. 61, two things are striking:

- First the different degradation of $E_\perp$ and $G_{\perp\parallel}$
- and second, the difference between GFRP and CFRP.

Both results are reproducible and go in line with FEM-results and damage mechanical modeling. Recent studies [Lambrecht 2007] confirm that with the degradation curves of Fig. 61 a realistic post failure analysis is possible for practically all laminate set ups and load cases. The study confirms, too, that the major Poisson's ratio is not degraded by IFF and that consequently η_ν in (Eq. 118) should be set to "1".

5.2.4 Degradation method used in the "World Wide Failure Exercise"

For the materials used in the World Wide Failure Exercise [Soden et al. 1998] no information about the degradation of moduli after IFF was given by the organizers. Therefore, Puck and Schürmann [Puck and Schürmann 1998] decided to use an alternative approach instead of Puck's method of 1969. This alternative approach does not use experimentally determined parameters. Instead Puck and Schürmann kept the stress exposure after IFF-onset constant at $f_E = 1$ by degrading the moduli correspondently. This method works quite well. However, the definition to keep the stress exposure constant at $f_E = 1$ is a reasonable, but after all arbitrary assumption. After all, at the time of the World Wide Failure Exercise No 1, there was no knowledge about the necessity to degrade $E_\perp$ and $G_{\perp\|}$ very differently.

5.2.5 New degradation method (2007)

After the publication of the experimental results of the author [Knops 2003] (compare chapter "Experimental determination of degradation curves"), which show that $E_\perp$ and $G_{\perp\|}$ have to be degraded very differently, it became even more desirable to calculate the stresses after IFF not with the fixed moduli which were valid at the onset of cracking. Puck made an improved proposal compared to the $f_E = \text{const.} = 1$ method by decreasing f_E depending on the already reached medium crack density δ in the layer. For the degradation a linear function is used:

$$f_E = 1 - \alpha_{f_E} \cdot \delta \qquad \text{(Eq. 120)}$$

For the degraded moduli, linear relations are used, too:

$$E_{\perp s} = E_{\perp 0}\left(1 - \alpha_E \cdot \delta\right) \qquad \text{(Eq. 121)}$$

$$G_{\perp\| s} = G_{\perp\| 0}\left(1 - \alpha_G \cdot \delta\right) \qquad \text{(Eq. 122)}$$

The coefficients α_{f_E}, α_E and α_G allow the calibration with experimental data.

$E_{\perp 0}$ and $G_{\perp\| 0}$ are secant moduli taken from the $(\sigma_2, \varepsilon_2)$-curve or the (τ_{21}, γ_{21})-curve, respectively. The fundamental idea is to treat a cracked layer as a damaged material with degraded moduli and a degraded sustainable stress exposure f_E, expressed for instance by $\alpha_{f_E} = 0.5$. Different α_E and α_G can be used to degrade $E_\perp$ and $G_{\perp\parallel}$ differently. The simple linear functions make it possible to calculate the crack density for a loadstep with a quadratic equation. This leads to over all low computational effort. The new method will shortly be published as [Puck and Deuschle 2008].

5.2.6 Laminates with unsound loading conditions

Experienced designers know that a laminate which in service has to withstand different load combinations should be designed with at least three different fiber directions. Laminates having only two fiber directions can also be exposed to a "reasonable loading condition". An example is a ±45°-tube where a tortional load results in a pure τ_{xy}-stress. For a ±55°-tube with closed ends internal pressure may be regarded as a reasonable load case, too. However, the real stress/strain- and fracture behavior of such a pressure vessel is quite complicated (compare Chapter "Validation of degradation curves") and it is recommended to use a third fiber direction for this case, too.

A very simple check for finding out about laminates with "unsound loading conditions" is the application of netting analysis to the laminate. The "±55°-tube" results for instance from applying netting analysis to the case of a pressure pipe with a stress ratio $\sigma_y{:}\sigma_x = 2{:}1$. The only possible fiber direction for a fiber network with just 2 directions is here $\alpha = \pm\arctan\sqrt{2} \approx \pm 54.74°$. Obviously a pure fiber network with α deviating from ±54.74° cannot provide equilibrium. The other way round, netting analysis tells that the fibers alone are not able to carry loads which belong to stress ratios $\sigma_y{:}\sigma_x$ deviating from exactly 2:1, even if the deviation is extremely small. If a ±54.74°-tube is tested with a stress ratio which considerably deviates from 2:1 after IFF has occurred, the forces in the laminate cannot concentrate in the fibers as it is intended for fiber reinforced components. Equilibrium between external loads and internal forces can only be found with undesirably high stresses σ_2 and τ_{21} attacking the matrix and the fiber matrix interface. This leads to unusually high nonlinearity of the stress/strain curves and to heavy matrix cracking immediately after an early onset of cracking (compare Chapter "Validation of degradation curves"). At very low stresses the laminate will loose more and more of its original integrity and it will not recover from these damages when it is unloaded. This happens because after crack onset there is no pronounced concentration of the forces in the fibers and the laminate undergoes large deformations in order to produce the stresses which are necessary for equilibrium with the external loads. Very soon such a high degree of deterioration of the composite can be reached that the tube can no longer be used even under its normal service condition $\sigma_y:\sigma_x = 2:1$. In this sense the laminate may have "failed" long before the maximum load (measured in a strength test) has been reached.

Puck stopped in his work for the World Wide Failure Exercise the calculations for such a component at a shear strain $\gamma_{21} = 0.2$ [Puck and Schürmann 2002] which is more or less an arbitrarily chosen limit. Certainly, it is not based on a physical consideration, because there is no known correlation between excessive shear deformation and deterioration of the laminate.

A similar undesirable situation is found for instance for a ±45°-tube with a stress ratio $\sigma_y:\sigma_x = 1:-1$. This leads to the same loading condition as found in a [0°/90°] laminate under a pure shear stress τ_{xy}, meaning $\tau_{\perp\parallel}$-stressing for all laminae. The fibers cannot pick up considerable longitudinal forces, in fact they are stressed nearly by the same low shear stress the matrix is stressed with. The individual layers of the laminate are exposed to the same longitudinal shear stress τ_{21} as an isolated UD-layer under pure τ_{21} shear stress, at least as long as no IFF have occurred.

After IFF onset the cracks are locally bridged to a certain extend by the fibers of the neighboring layers. Therefore, the shear stress τ_{21} at total failure of the ±45°-tube (with a stress ratio $\sigma_y:\sigma_x = 1:-1$) is a little higher than the longitudinal shear strength $R_{\perp\parallel}$ of the lamina. However, after $\tau_{21} = R_{\perp\parallel}$ has been exceeded, the laminate will very soon become deteriorated and useless.

The conclusions to be drawn from the facts discussed here are the following:

- In practice laminates as discussed here can only be used if nothing but very small deviations from those loading conditions occur which for the netting analysis finds equilibrium of loads and forces.
- It is extremely difficult to perform a reasonable "failure" analysis for laminates with "unsound loading conditions", because there is no clear definition of "failure" in this case.

5.3 Implementation of the Puck theory for the gradual failure process in a software code

The basis for the use of the Puck theory is a stress analysis which takes all non-linear effects into account (compare Chapter "Nonlinear stress analysis before IFF" in the Annex). Additional elements of the analysis are

Puck's fracture criteria, a realistic procedure for the degradation of elastic properties after damage (IFF) initiation and – last but not least – a correct exit condition which defines what is regarded as ultimate laminate failure.

In the following the numerical procedure which is needed for the complete implementation of Puck's theory is presented in detail. The comprehensive description is especially useful for readers who intend to program Puck's theory themselves.

In detail, the stress analysis should take into account residual stresses and in addition the following non-linear effects:

- The non-linear stress/strain-behavior of the lamina before IFF
- The non-linear stress/strain-behavior of the laminate caused by IFF (i. e. the gradual failure process)
- The change in fiber direction of the lamina under load.

A database for the non-linear behavior can – for instance be taken from Soden and Kopp [Soden et al. 1998, Kopp 2000] including non-linear stress/strain-curves ($\sigma_2(\varepsilon_2)$ and $\tau_{21}(\gamma_{21})$) for GFRP and CFRP.

Often the stress analysis evaluates only 2D-stresses and is based on the Classical Laminate Theory. In this case the 2D-Puck criteria are used for the fracture analysis. If however, the stress analysis is 3D then the formulae summarized in Fig. 42 are used for the search of the fracture plane and the IFF-limit.

The gradual failure process can be programmed according to any of the models presented in the chapter "Puck's approach for the analysis of the gradual failure process". The procedure described below fits, however, to the original Puck model from 1969.

The load should be applied to the laminate in a discrete number of load-steps (at least 50 to 100 steps). The analysis for the first step has to be done with the starting moduli ($E_\perp$ and $G_{\perp\parallel}$) and the pre-defined fiber-directions. First the laminate and laminae strains, the resulting stresses and the change in fiber direction are calculated (compare (Eq. 117) (in the Annex)). The fracture analysis is then done with the Puck criteria. $E_\perp$ and $G_{\perp\parallel}$ are recalculated for the next load step according to the stored non-linear $\sigma_2(\varepsilon_2)$- and $\tau_{21}(\gamma_{21})$-curves. As long as no IFF occurs this recalculation can be limited to $E_\perp^c$ and $G_{\perp\parallel}$.

The shear modulus $G_{\perp\parallel}$ can directly be taken from the (τ_{21}, γ_{21})-curve of the UD-lamina, whereas for the determination of $E_\perp^c$ first the combined strain ε_2^{comb} has to be calculated (compare (Eq. 115) and (Eq. 116)) and the Young's modulus is then taken at ε_2^{comb} from the (σ_2, ε_2)-curve.

Subsequently, the laminate is – with the adapted stiffness matrix – loaded with the increased load (next load-step). The accuracy of the results is – if this mathematically simple procedure is chosen – to some extent depending on the step size. However, if the number of load steps is chosen reasonably large (> 10 before IFF initiation) this dependence is negligible.

The algorithm after IFF-onset in at least one lamina is in principle the same as it is before IFF-onset. The only difference is that instead of the non-linearities due to micro-damage the non-linearities due to IFF are considered for the calculation of the elastic properties for the next load step, too. The procedure is explained in detail in the following.

The IFF-cracked lamina is still considered as a continuum with degraded elastic properties according to (Eq. 118). In (Eq. 118) $E_{\perp s}$ and $G_{\perp\parallel s}$ and $\nu_{\perp\parallel s}$ ($\nu_{\perp\parallel}$ is not degraded according to the experimental and numerical results presented in chapter "Experimental determination of degradation curves") are the moduli at the IFF-fracture limit. The parameters for the degradation are taken from Table 4, but other parameter sets can be defined by the user, too.

The stresses after the first IFF in a lamina (not the strains of the laminate!) are calculated with the strains (ε_1, ε_2, γ_{21}), the Young's modulus $E_{\perp s}$ and the shear modulus $G_{\perp\parallel s}$ (incorporating only the non-linearities up to IFF-initiation and not the degradation due to IFF). Thus these stresses are higher than the corresponding stresses at the IFF-limit. These stresses are called "fictive" stresses because they are not really present in the lamina. With the fictive stresses the stress exposure $f_{E,\,IFF}$ is calculated – leading to values $f_{E,\,IFF} > 1$.

The fictive values are only needed for the determination of the degradation parameters η_E and η_G. In addition to the fictive stresses "real" stresses – averaged over the cracked lamina – are calculated with the degraded properties $E_\perp(\eta_E)$ and $G_{\perp\parallel}(\eta_G)$ and the strains (ε_1, ε_2, γ_{21}) of the lamina. These real stresses are used for the determination of a "real" averaged stress exposure f_E (smaller than the fictive value), the fracture mode and the fracture angle.

This procedure is necessary to use on the one hand the stress exposure as control parameter for the degradation and to calculate on the other hand the fracture mode and the fracture angle – which both might change with increasing load from one crack to the next because of unequal degradation of $E_\perp$ and $G_{\perp\parallel}$ or (load dependent) change in fiber direction.

If in a loadstep a stress exposure is calculated which is equal or lower than in the preceeding loadstep, then the moduli have to be taken from the preceeding loadstep and kept constant.

The above procedure has for instance been realized in the software AlfaLam.nl (download from the homepage of the institute "klub" of the TU Darmstadt: http://www.klub.tu-darmstadt.de/forschung/download.php) and the scientific software *Subu*, which has been developed by the author [Knops 2003]. An additional detail of this program is the following: The non-linear stress/strain-curves (Fig. 95) are only used until the first IFF occurs, which happens under combined ($\sigma_\perp$, $\tau_{\perp\parallel}$)-stressing at considerably lower strains $\varepsilon_\perp$ and $\gamma_{\perp\parallel}$ than under uniaxial stressing. The background of this procedure is that all non-linearities which occur after IFF-initiation are considered in the η-curves. As mentioned above, the ratio of the stresses σ_2/τ_{21} can change in the course of the load application due to the different degradation curves of $E_\perp$ and $G_{\perp\parallel}$ and the change in fiber direction. This can lead to a change of the fracture mode and the fracture angle. These effects are correctly calculated by *Subu*. The flowchart of the program *Subu* is shown in Fig. 62.

5.4 Summary of chapter

Inter Fiber Fracture is sometimes regarded as final fracture and in other cases as tolerable damage of the laminate. If IFF is regarded as tolerable, a "post IFF-analysis" of the laminate is necessary which comprises as an iterative process strain and stress analysis with elastic properties which are truly characteristic for the damaged laminate. This post IFF-analysis is understood here as "Gradual Failure Analysis". The different approaches for the gradual failure analysis (fracture mechanics, damage mechanics and phenomenological models) are discussed briefly at the beginning of the chapter. Puck's degradation models are then presented in more detail. Finally the implementation of Puck's theory in a software code is described in detail.

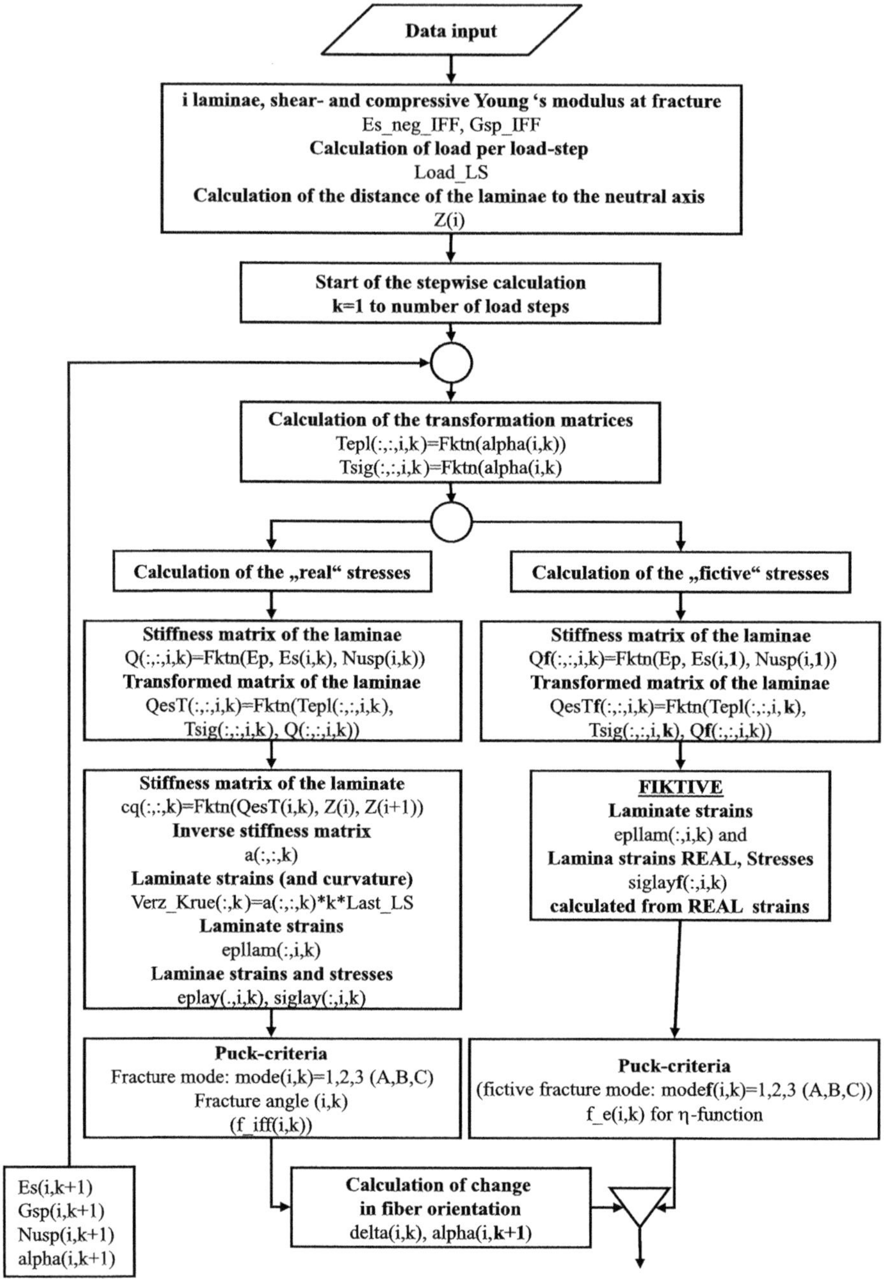

Fig. 62. Flowchart of the program Subu

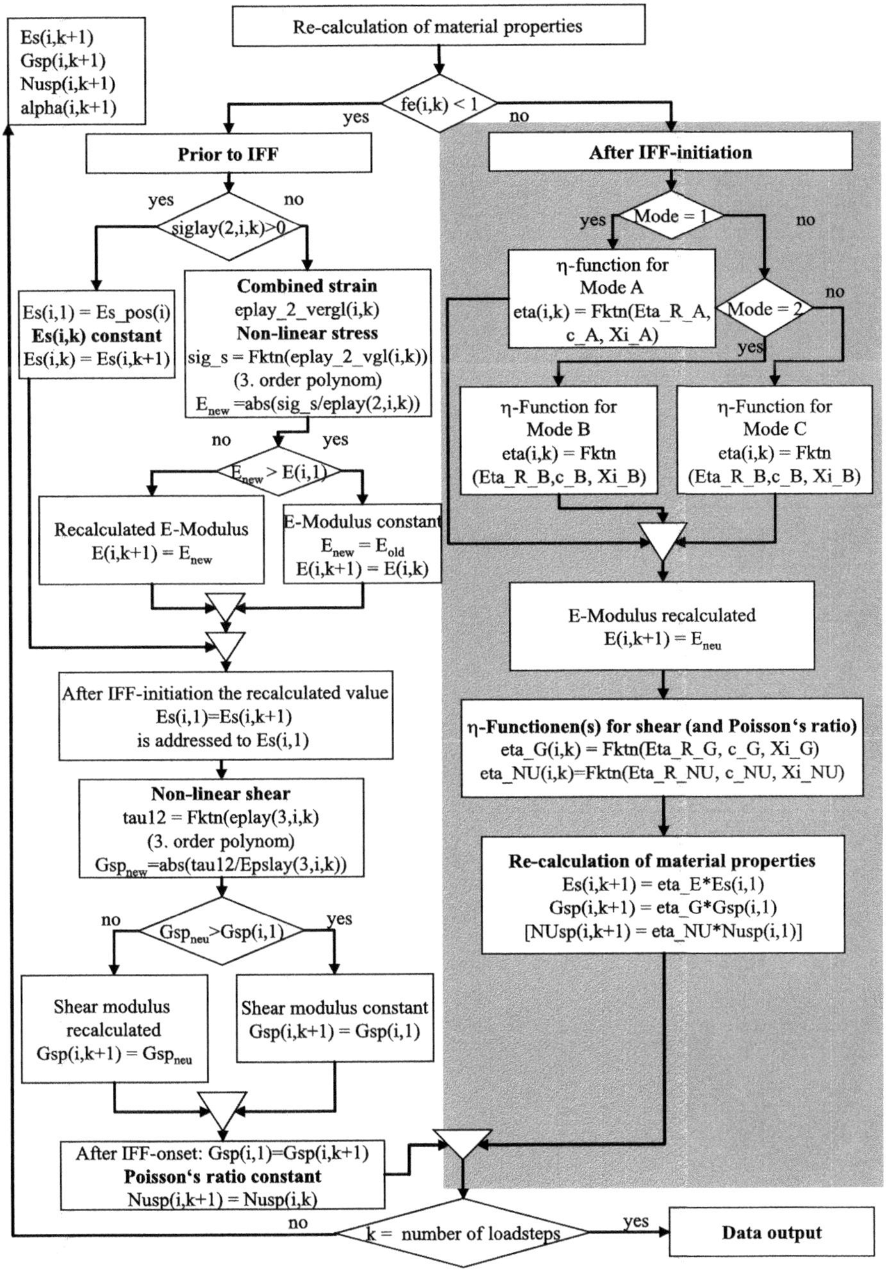

Fig. 62. Continued

6 Experimental work

6.1 Verification of the fracture hypotheses and calibration of the fracture body

When the work on the experimental verification of the Puck criteria started in the mid 1990s in Germany, none of the common strength criteria was adequately verified. The reason for this is that fracture experiments with combined stressing are extraordinary difficult to realize and very costly. Nevertheless in many places experience was gathered, which clearly demonstrates, that FF and IFF are of completely different nature not allowing to mix them in one strength criterion, like it is done in global criteria.

Beside the correct depiction of fracture data the most important hint concerning the validity of the action plane related criteria is the experimental confirmation of theoretically predicted fracture angles θ_{fp}. To answer this question intensive experimental work was carried out in a series of doctoral theses and documented in [Huybrechts 1996, Cuntze et al. 1997, Kopp 2000]. A comprehensive summary of the work can be found in a SAMPE publication from 1998 [Kopp Michaeli 1998]. If not mentioned explicitly the material used for the specimens is glass/epoxy with LY556/HY917/DY070 as epoxy system. Glass fibers were chosen because of the translucency of GFRP.

6.1.1 Experiments with Three Dimensional Stress States

To verify the fracture angles calculated by the Puck criteria for 3-D-stress states, experiments on UD-laminae with $\tau_{\perp\perp}$, $\sigma_{\perp}$ and $\tau_{\perp\|}$ stressing are necessary.

Such test configurations however are difficult to realize and it took years to finalize that task to full satisfaction.

6.1.1.1 Beam specimens for (τ_{23}, σ_2) stress combinations

The starting point was in 1995 a 3-point-bending test. With this test concept, stress combinations of τ_{23} and σ_2 can be realized on the UD-lamina. In Fig. 63 the experimental set-up originally used by Huybrechts [Huybrechts 1996]

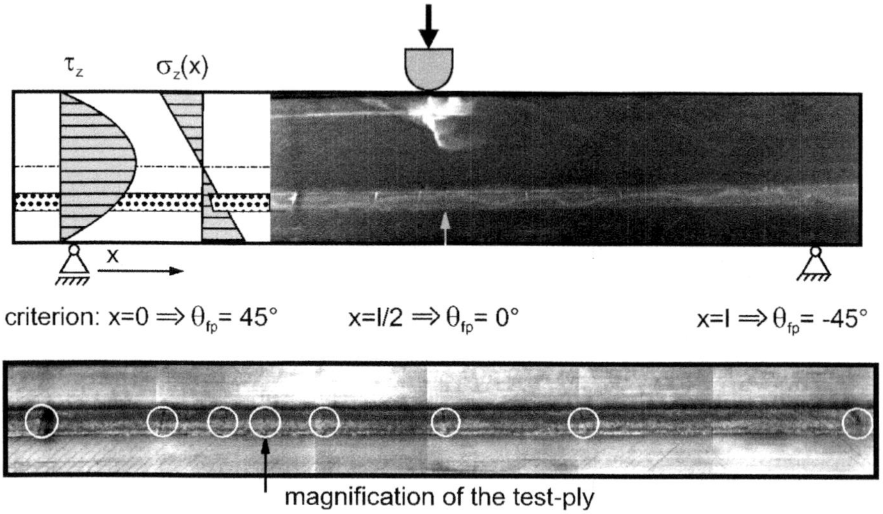

Fig. 63. Layered three point bending test

is shown. The specimens consist of stiff supporting layers with 0°-fiber direction on top and bottom and a 90°-test layer which is embedded in the 0°-layers.

Figure 64 shows the results of an FEM-stress analysis and the optimized specimen's geometry [Kopp 2000]. The 0°-layers on top and bottom are made of CFRP and for the test layer of GFRP. The layers were cured separately and later bonded together to avoid thermal stresses. The use of CFRP for the supporting layers and GFRP for the test layer guaranties a homogenous transverse contraction of the layers and avoids stresses due to different Poisson ratios of the layers.

In the specimens IFF develops in the 90°-test-layer at relatively low load levels, which do not endanger the integrity of the complete test specimen. Different states of stress are realized by varying the test-layer's position over the specimen thickness. If the test-layer is positioned in the tension zone of the beam, the stress profile in the test layer changes over the beams length from a pure transverse/transverse shear stress τ_{23} right above the support to a stress state with more and more transverse tension σ_2^t (compare Fig. 63). Consequently, the maximum transverse tension is reached exactly beneath the load introduction. According to the Puck criterion the fracture angle has to change from θ_{fp} = 45° above the supports (due to pure $\tau_{\perp\perp}$ stressing, compare Fig. 65) to an angle perpendicular to the beam axis beneath the load introduction.

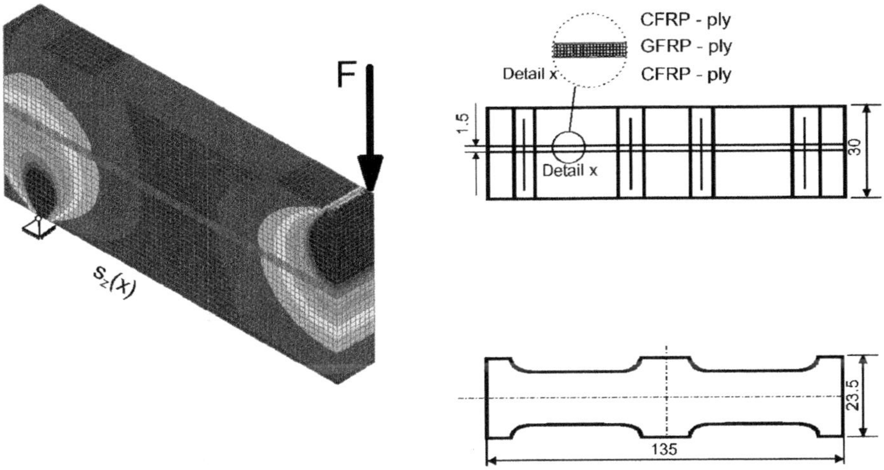

Fig. 64. FEM-Analysis of a beam specimen and optimized geometry

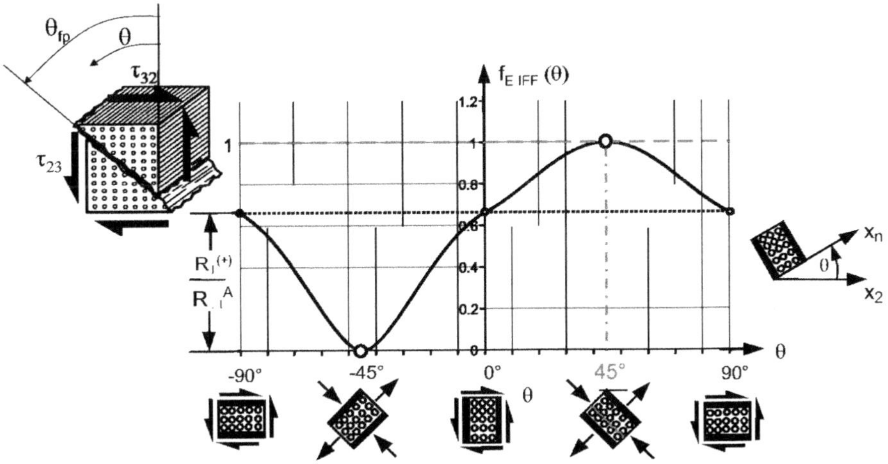

Fig. 65. Angle dependent stress exposure for pure transverse/transverse shear stressing

A disadvantage of the 3-point bending tests is the small test zone due to the limited thickness of the test-layer. The test layer is thin, because in a thick layer the homogeneity of the stress state is not given.

The beam specimen tests validated qualitavely the fracture angles calculated with the Puck criterion. However, due to the described shortcomings of the beam specimen the latter has not been further developed but alternative test set ups were investigated.

6.1.1.2 Torsional ring specimens for (τ_{32}, τ_{31}) stress combinations

Another specimen concept used for the investigations is that of a torsional ring (Fig. 66). With this concept different shear stress combinations (τ_{32}, τ_{31}) corresponding to a combined ($\tau_{\perp\perp}$, $\tau_{\perp\parallel}$) stressing can be realized. At the inner and outer shell surface of the GFRP test ring oppositely directed torsion moments of equal magnitude are introduced by means of metal gear-wheels. Disturbing stresses were excluded and an absolutely homogenous shear stressing over the radial extension was obtained by a finite element geometry optimization resulting in a tailored cross section of the GFRP ring. Depending on the test layer's fiber direction with respect to the axial direction of the ring all stress combinations between a pure transverse/transverse shear stress τ_{32} corresponding to $\tau_{\perp\perp}$ stressing (axial fiber direction) and a pure transverse/longitudinal shear stress τ_{31} corresponding to $\tau_{\perp\parallel}$ stressing (circumferential fiber direction) can be realized.

A lot of work was necessary to optimize the test concept and the geometry of the specimens. The load is introduced into the specimen via gear-wheels on the inner and outer side. The GFRP-rings are bonded to the metal gear-wheels. The geometry of the ring has been optimized with the goal of keeping the stresses and the stress exposure constant in the testing zone. The geometry and the homogeneous distribution of the stress expo-

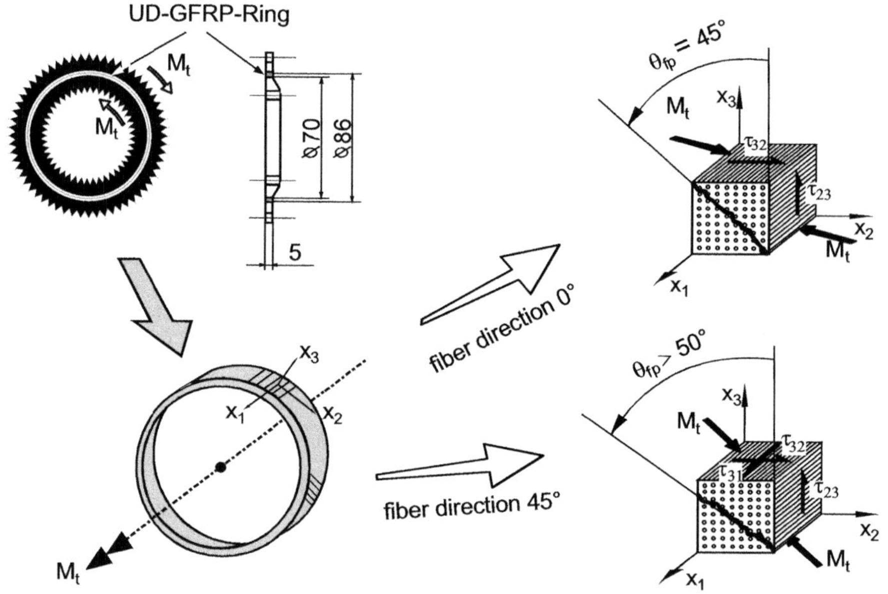

Fig. 66. Torsion ring test specimen [Kopp Michaeli 1998]

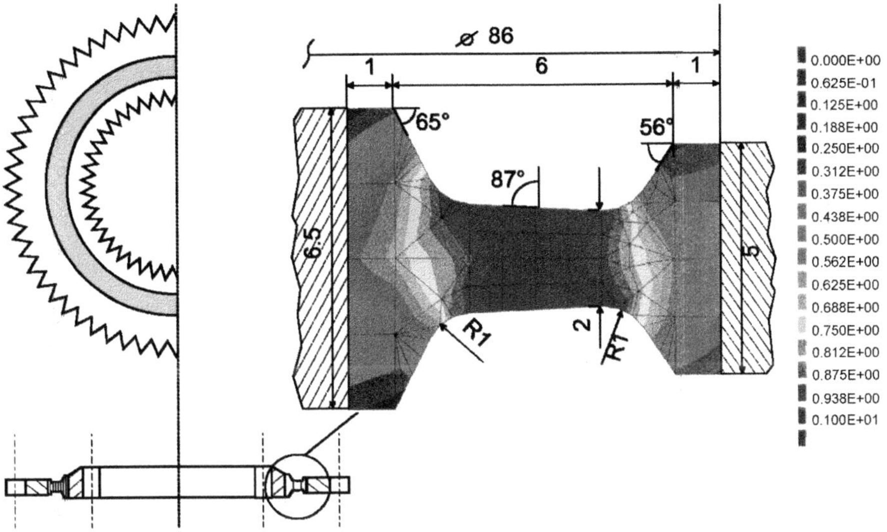

Fig. 67. Optimized torsion ring cross section and stress exposure $f_E(\theta_{fp})$

sure over the test zone are shown in Fig. 67. The test zone is approximately 5 mm wide, the cross section of the test zone is turning machined from the filament wound ring.

As an example for the whole test series with fiber directions ranging from $\alpha = 15°$ to $\alpha = 90°$ in 15° steps ($\alpha = 0°$ corresponds to the direction of the ring axis), the test results for a fiber direction of $\alpha = 15°$ are presented in the left part of Fig. 68. The stress state in the test ring is in this case close to a pure $\tau_{\perp\perp}$-stressing of the UD-lamina. Accordingly the fracture angle is predicted to be close to $\theta_{fp} = 45°$. The statistical evaluation of the 150 almost normally distributed IFF fracture angles validates the Puck model for pure $\tau_{\perp\perp}$-stressing.

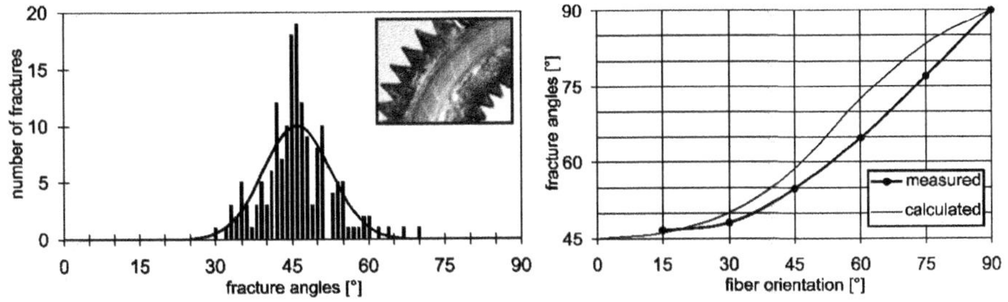

Fig. 68. Results of the torsional ring tests

Table 5. Evaluation of the torsion ring test specimens [Kopp2000]

α	$\tau_{\perp\parallel}/\tau_{\perp\perp}$	Number of specimens tested	Number of IFF	Fracture angle θ_{fp}
15°	0.27	23	266	47°± 5.5°
30°	0.58	10	49	49° ± 7°
45°	1	11	24	55° ± 5.5°
60°	1.73	10	33	65° ± 12.5°
75°	3.73	10	25	77° ± 6°
90°	∞	11	4[22]	90° ± 0°

The right picture in Fig. 68 compares the entire evaluation of the test series with the theoretically predicted fracture angles θ_{fp}. The correspondence between predicted and measured values is very satisfying. The experimental results (mean and standard deviation of the fracture angle θ_{fp}) of Kopp are summarized in Table 5.

6.1.1.3 Investigation on a damage threshold due to micro cracking under compressive (σ_2^c, σ_3^c) stress combinations

One aspect of Puck's IFF criteria is the unclosed Master Fracture Body (compare Fig. 41, Fig. 45, Fig. 56) resulting from the second part of hypotheses 2. This leads to an unclosed (σ_2, σ_3)-fracture curve (Fig. 69) in the third quadrant just like the one of Hashin (compare Fig. 55), which may not be interpreted in the way that the unidirectional lamina is able to withstand any desired loading as long as $\sigma_2 < 0$ is equal to $\sigma_3 < 0$.

Kopp [Kopp 2000] conducted a sophisticated micromechanical analysis and backed the analytical results with hydrostatic compression tests which induce the desired (σ_2^c, σ_3^c) stress state, where $\sigma_2 = \sigma_3$. The result of this analysis is summarized in Fig. 70. This diagram can be interpreted the way that at roughly a stress level of $\sigma_2 = \sigma_3 = -2R_\perp^c$ micro cracking in the matrix has reached a level which has a significant influence on the transverse tensile strength of the UD-lamina.

[22] 7 torsion rings failed in the bond line. Due to the excellent reproducibility of this test series (see also θ_{fp} for tubular specimen in [Cuntze et al. 1997]) no additional tests were performed.

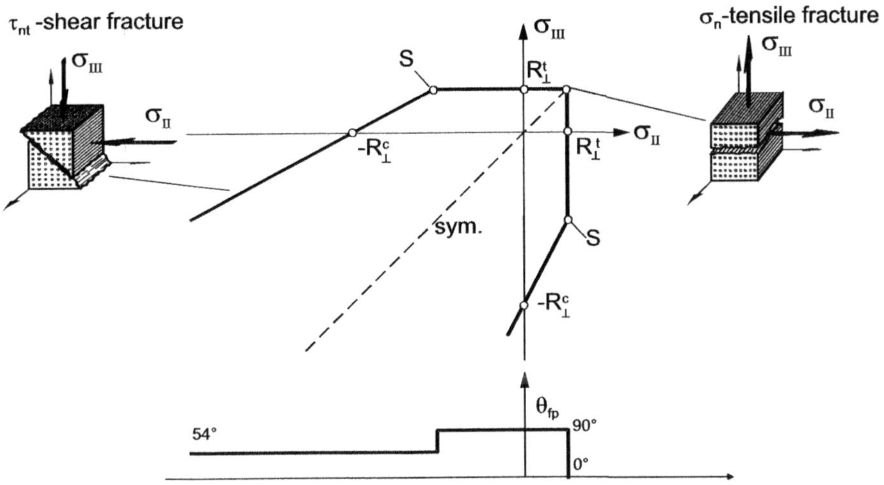

Fig. 69. (σ_2, σ_3) fracture envelope

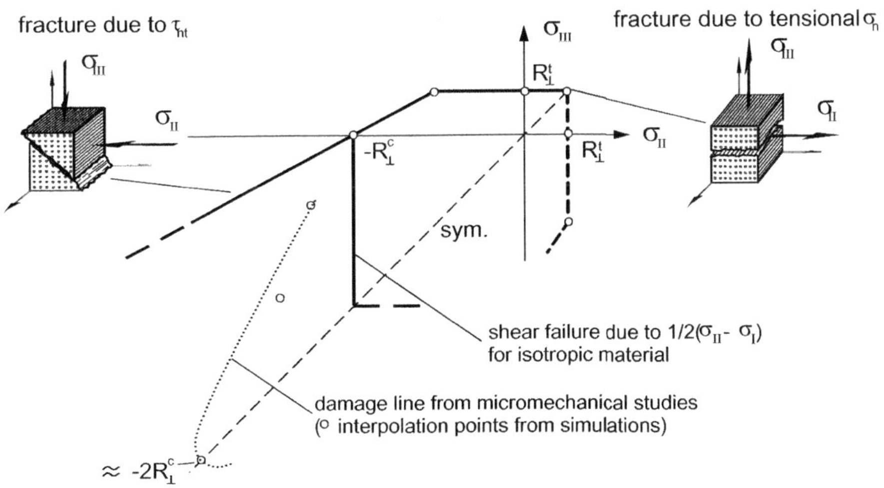

Fig. 70. (σ_{II}, σ_{III}) fracture curve and damage threshold from micro-mechanical analysis and hydrostatic tests

6.1.2 Experiments with plane (σ_2, τ_{21})-stress combinations

For most composite structures the assumption of a plane (σ_1, σ_2, τ_{21}) state of stress is reasonable. Thus, the validation of Puck's failure criteria for this most common case is of major importance. The plane state of stress is

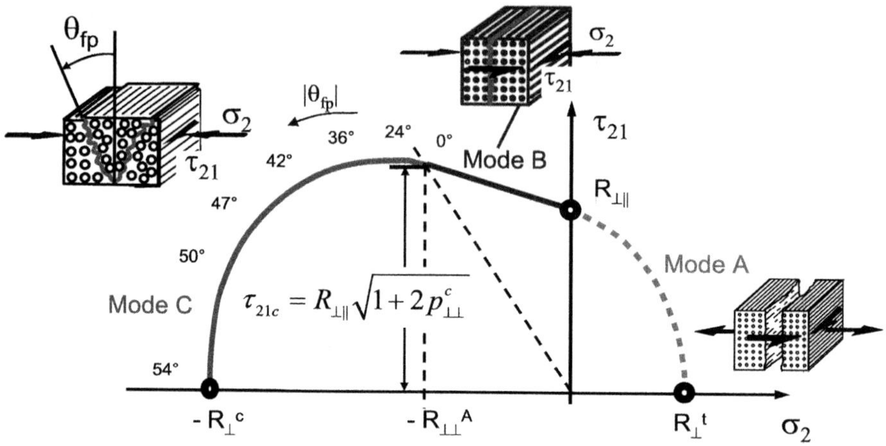

Fig. 71. (σ_2, τ_{21}) fracture curve, fracture modes and angles

of course a special case covered by the universal 3D-Puck-criteria. However, in this case the searching procedure for the fracture plane with its iterative computation for all fracture angles θ between –90° and 90° is not required as direct analytical solutions have been found by *Puck*, which can be handled as easily as the traditional fracture criteria (compare Chapter Analytical 2-D-formulation for plane states of stress).

In Fig. 71 the IFF-fracture curve is shown as it is calculated by (Eq. 71), (Eq. 73) and (Eq. 77). As explained in the Chapter 4.2.7 this curve is subdivided in 3 parts corresponding to fracture Modes A, B and C. The verification of the fracture angle for Mode A ($\theta_{fp} = 0°$) is straight forward and has – among others – already been published in [Puck 1967, Nairn and Hu 1994, Huybrechts 1996]. Of more interest is the clarification of the fracture limits and the fracture angle in the regime of Mode B (is it really $\theta_{fp} = 0°$ as for Mode A) and above all the verification of the slope of the fracture angle in Mode C rising – according to Puck – continuously from 0° to ±54° with increasing transverse compression.

6.1.2.1 Experiments with tubular specimens to verify the fracture limits

The verification of the fracture limits is easier than that of the fracture angles and was conducted first. The test data in Fig. 72 represent an overview of the fracture stress combinations on nine different load paths determined in the experimental campaign documented in [Cuntze et al. 1997]. For these

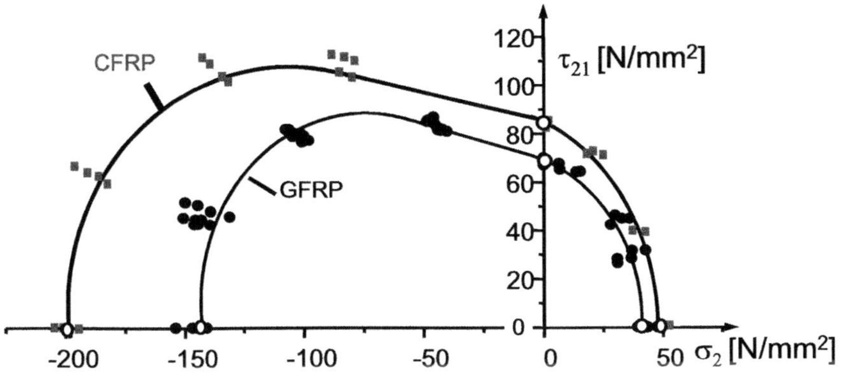

Fig. 72. (σ_2, τ_{21}) fracture curve, fracture limit from experiments

fracture tests filament wound tubular specimens made of glass epoxy, carbon epoxy respectively were used and tested on a Torsion/Tension/Compression machine. This machine can apply simultaneously torsion and tension (compression respectively) in axial direction of the tubular specimens. The action plane related IFF criterion maps the experimental data excellently, see also [Hinton et al. 2002, Soden et al. 2002], too.

6.1.2.2 Experiments with slant shear specimens to verify the fracture angle for (σ_2^c, τ_{21})-combinations

The evaluation of the fracture angle in the regime of Mode B and C is much more complicated than that of the fracture limit and practically not possible with thin walled tubular specimens. Therefore, Kopp optimized for the verification of the fracture angle in the regime of Modes B and C the so called Slant Shear test setup of Kadatoni and Aki [Kadatoni and Aki 1984]. Figure 73 shows the principle of this test and some preliminary results achieved by Kopp. The experimental setup offers the opportunity to vary the shear stressing $\tau_{\perp\|}$ superimposed on the compressive stressing $\sigma_\perp^c$ by changing the fiber direction or the inclination of the laminated specimen embedded in the metal load introduction units.

Further optimization by FEM led to tailored full composite specimens without metal load introduction as shown in Fig. 74. There are no stresses resulting from the different Poisson ratios of steel and the laminate in the test zone of the specimens. Besides, it is secured that the highest stress exposure occurs always in the testing zone. The fracture does not originate from a free edge or an area outside the testing zone.

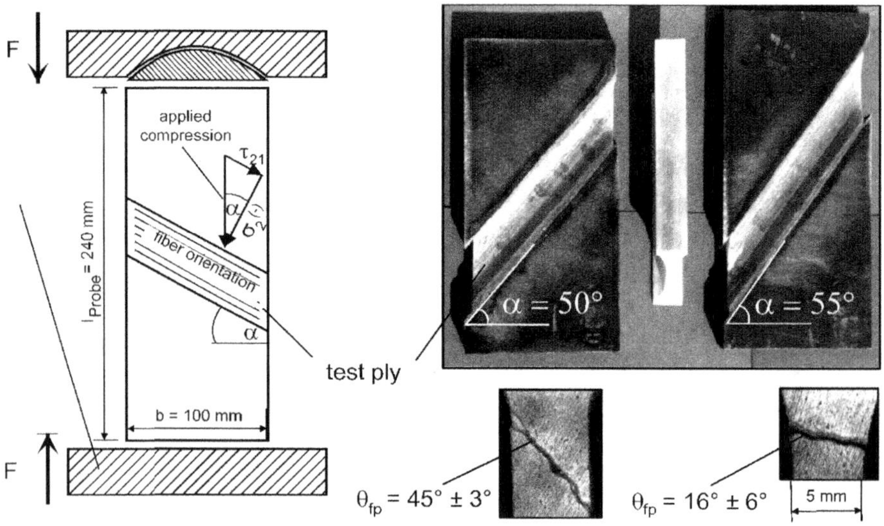

Fig. 73. Slant shear specimens and preliminary tests

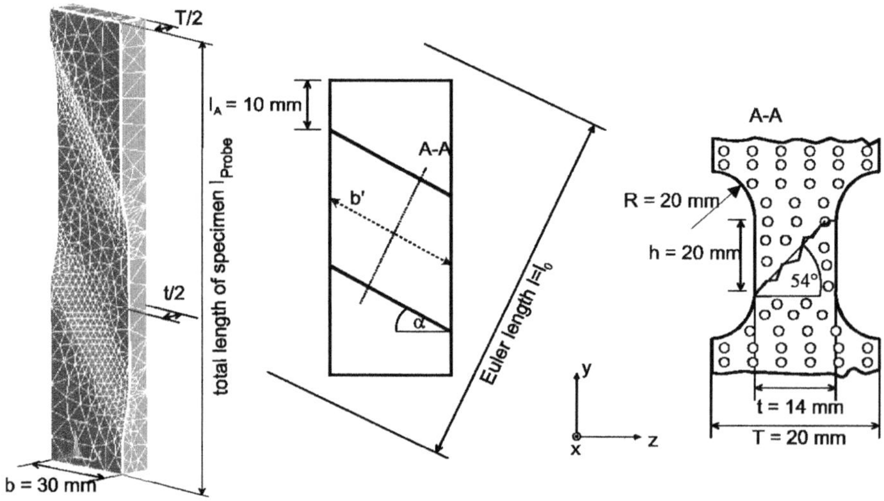

Fig. 74. Optimized slant shear specimens [Kopp2000]

The specimens were filament wound on a special tool (compare Fig. 75). After winding the laminate is pressed to a predefined thickness of 20 mm to arrive at a homogeneous fiber volume fraction of φ = 60 %. Similar to the torsional rings the specimens were sealed with a clear gel-coat after the grinding to the final geometry.

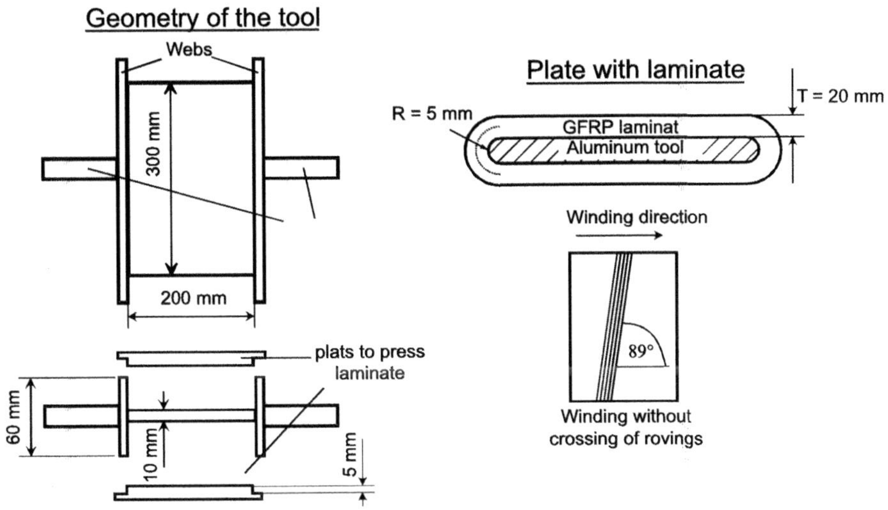

Fig. 75. Filament winding with non-crossing rovings [Kopp 2000]

Table 6. Evaluation of the slant shear tests [Kopp 2000]

α	$\lvert\sigma_\perp^c\rvert/\tau_{\perp\parallel}$	Number of specimens	Stress at fracture $\lvert\sigma_\perp\rvert$	Fracture angle θ_{fp}
65°	0.47	7	31 ± 1.5 N/mm^2	$11° \pm 8°$
60°	0.58	7	39 ± 1 N/mm^2	$14° \pm 5°$
55°	0.7	5	47 ± 1 N/mm^2	$20° \pm 4°$
50°	0.84	4	56 ± 0.5 N/mm^2	$34° \pm 6°$
45°	1	4	61 ± 2 N/mm^2	$43° \pm 3°$

Table 6 summarizes the results of all the slant shear tests conducted. Figure Fig. 76 compares these results to the theoretical values predicted by the Puck criterion.

The measured fracture angles match for the whole test series of several $\lvert\sigma_2^c\rvert/\tau_{21}$ ratios very well the predicted values. On average they are just slightly higher than those calculated by means of (Eq. 77). All in all the whole test series proves impressively that the fracture angle continuously increases with growing ratio $\lvert\sigma_\perp^c\rvert/\tau_{\perp\parallel}$. Thus the second of Puck's fracture hypotheses (compare chapter 4.2.3) stating that a compressive stress σ_n^c impedes fracture on its fracture plane whereas τ_{nt} and τ_{n1} interact in causing the fracture has been successfully verified.

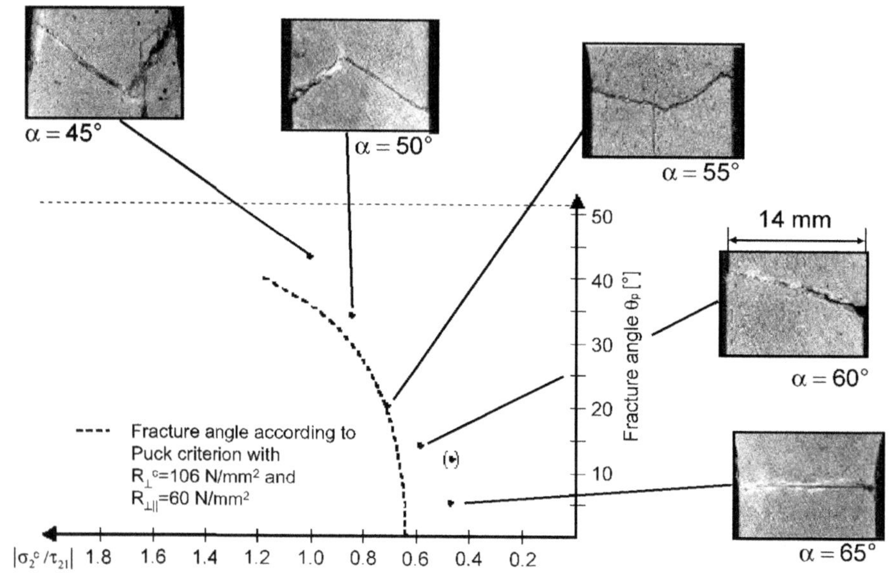

Fig. 76. Comparison of measured and postulated θ_{fp}

6.1.2.3 Experiments to determine the fracture angle under pure transverse compression

The transverse uniaxial compression test (pure compressive stress σ_2^c or σ_3^c respectively) is of great importance for the validation of the Puck criteria, namely of hypothesis 2 which states that a compressive stress impedes shear fracture on its action plane. Without this impeding effect the fracture would occur on the planes with the maximum shear stress ($\theta_{fp} = \pm 45°$). On these planes a compressive stress $\sigma_n^c = \sigma_\perp^c \cdot \cos^2 45°$ is acting in addition to the shear stress $\tau_{nt} = \sigma_\perp^c \cdot \sin 45° \cdot \cos 45°$ (compare (Eq. 1) and (Eq. 2)). Thus, under ±45° compressive stress and shear stress are of equal magnitude (Fig. 77).

According to Puck, the transverse compressive stress impedes the shear fracture. On a section plane with $|\theta| > 45°$ the shear stress τ_{nt} which causes the fracture only slightly decreases whereas the fracture impeding transverse compressive stress σ_n^c sharply decreases (compare Fig. 77). The goal of the experiments described below was to verify this material behavior.

The experimental investigation of the fracture angle under transverse compression was started by Huybrechts [Huybrechts 1996]. The specimens look similar to those for the slant shear tests (compare Fig. 73) The specimen's geometry was later optimized by Kopp [Kopp2000] with FEM to make sure, that fracture is initiated in the test zone. Figure 78 shows not

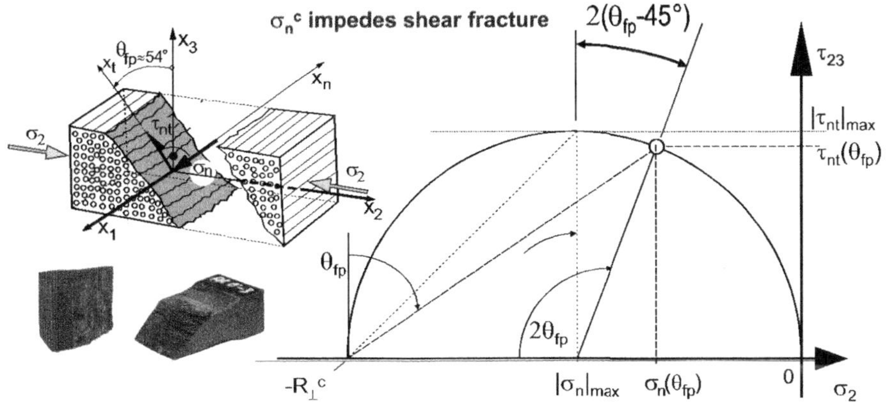

Fig. 77. Stresses on inclined sections due to pure transverse compression $\sigma_\perp^c$

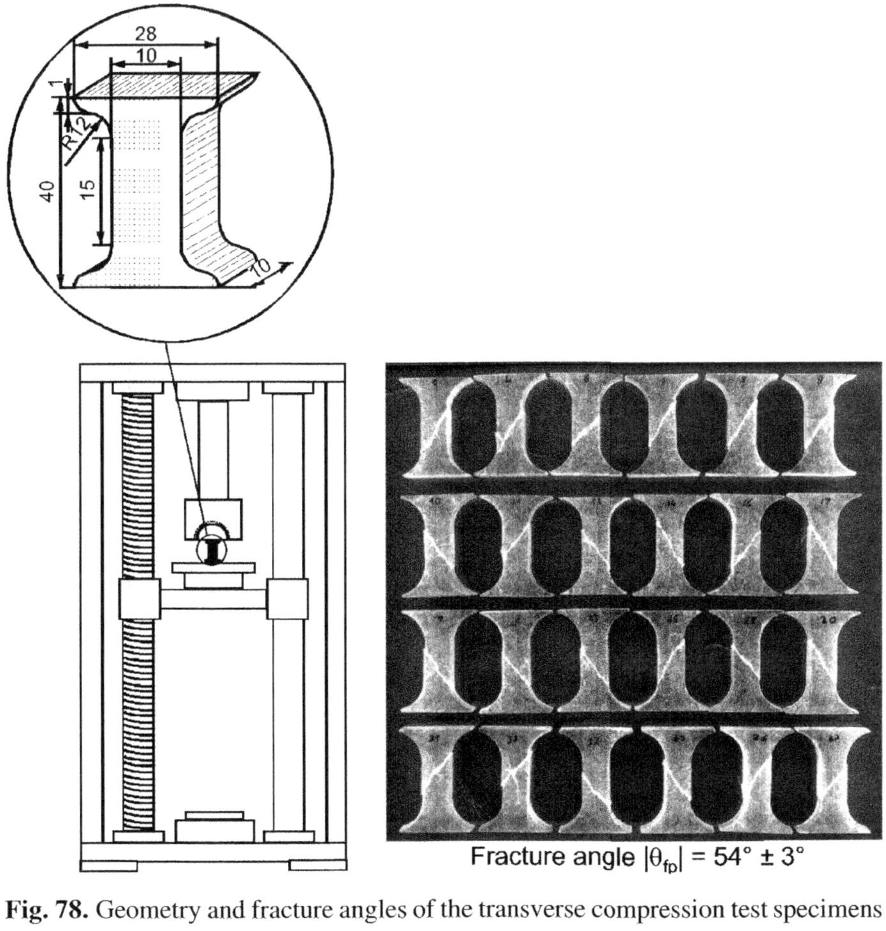

Fig. 78. Geometry and fracture angles of the transverse compression test specimens

only the optimized geometry and the test machine, but also a picture of 24 tested specimens. The fracture angle was measured to be 54° ± 3°. All fractures occur well in the test region and are not initiated in the curved section of the geometry.

The fracture impeding influence of a transverse stressing σ_n^c was proved impressively by the test campaign. The measured fracture angle of 54° matches perfectly with the angle measured for cast iron [Hodgkinson 1861, Johnson 1926]. This indicates strongly that more generally the fracture behavior of the UD-lamina under transverse load (load perpendicular to the fibers) corresponds to that of brittle isotropic materials.

6.2 Experimental determination of degradation curves

Most of the published experimental work concentrates on flat specimens [Highsmith and Reifsnider 1982, Flaggs and Kural 1982, Edge 1998, Garret and Bailey 1977, Groves et al. 1987, Lavoie and Morton 2000, Boniface et al. 1997, Surgeon et al. 1999, Smith et al 1998, Smith and Wood 1990, Tsai and Daniel 1992] with cross-ply-laminates $[0, 90]_s$ and tensile testing in 0°-direction. Generally the data has been recorded for GFRP, because in cross-ply-laminates made of CFRP the loss of stiffness due to IFF in the 90°-layers is too small to be analyzed adequately. Some data concerning the degradation of the shear modulus has been generated too [Tsai and Daniel 1992]. However, due to an inhomogeneous state of stress in the testing zone and inconsistencies concerning the evaluation of the tests, this data is not regarded as reliable (for a detailed discussion see [Knops 2003]).

The data available from literature is illustrated in Fig. 79. The degradation η in Fig. 79 is defined as the coefficient of the degraded stiffness parameter and the correspondent value before IFF ($\eta=E_\perp/E_{\perp 0}$, $\eta=G/G_0$ respectively). All data has been recorded as a function of the crack density δ defined as the number of IFF in a layer divided by the length of the testing zone and multiplied with the layer thickness. Thus, phenomenological models like that by Puck cannot be calibrated with these data because they do not formulate a correlation between η and the crack density.

Based on this situation the author conducted a major experimental work to get experimental data for the calibration of phenomenological degradation models. In full detail this work is documented in [Knops 2003]. An extract can be found in [Knops Bögle 2006]. Here the work is presented in a more condensed way.

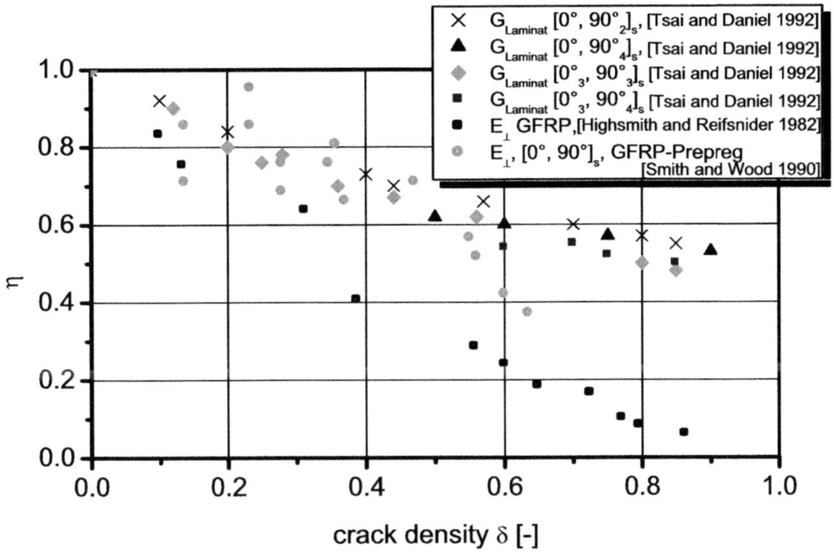

Fig. 79. Stiffness degradation data for GFRP-laminae from literature

6.2.1 Degradation of $E_\perp^t$ and $G_{\perp\parallel}$ for fracture Mode A

Both the degradation of the Young's Modulus transverse to the fibers $E_\perp$ and the degradation of the shear modulus $G_{\perp\parallel}$ were examined. Figure 80 illustrates the design of the flat specimens. The laminate is built up of a thick (5 mm or 10 mm respectively) 90° test layer being embedded in two thin (1 mm) CFRP supporting layers. The thick layers are manufactured on a special filament winding tool (compare Fig. 75), which was used by Kopp [Kopp2000], too. Thermal stresses are omitted by the separate curing of the three layers. After curing the layers are bonded together. Free transverse contraction of the test layer under uniaxial tension is guaranteed by adapting the Poisson's ratio of the supporting layers to the corresponding Poisson's ratio of the test layer. This was achieved by using CFRP-supporting layers made of woven fabrics with a warp/fill-ratio of 12:3 (warp in 0°-direction) in combination with GFRP test layers and UD-Prepregs with glass fibers in 0°-direction and HM-carbon fibers in 90°-direction in combination with CFRP-test-layers.

The tests were conducted on a hydraulic Tension/Compression-Torsion-Machine. The specimens were loaded with a monotonously increasing tensile force in 0°-direction. IFF was detected optically, acoustically and by the sudden drop of stiffness. The test was stopped right after the IFF occurred. The specimen was unloaded and reloaded again until a new IFF

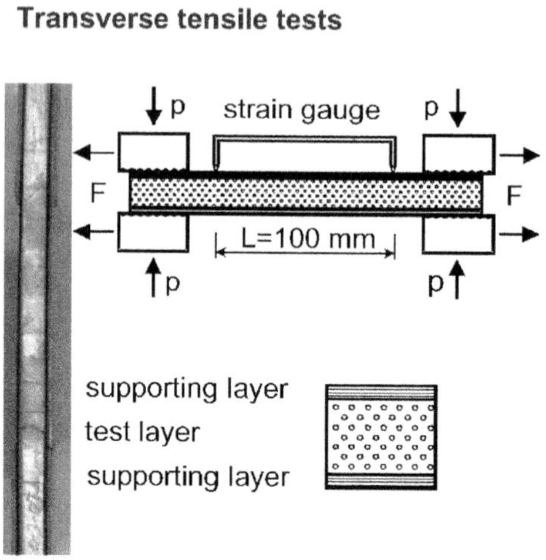

Fig. 80. Flat specimens to examine the degradation of $E_{\perp}$

occurred. This procedure was repeated until crack saturation was reached. Having measured the stress/strain-curve of the supporting layers separately, the stress/strain-curve of the test layer can be derived from the force/strain diagram of the specimen.

In Fig. 81 these curves are plotted for seven consecutive loadings of the specimen shown on the left side of Fig. 81. The numbers "1" to "6" indicate where the individual IFF occurred during the corresponding loading. All stress/strain-curves – with the exception of the last (7.) loading where delamination instead of further IFF occurred – are linear. Thus the tensile Young's Modulus $E_{\perp}$ can directly be derived from Fig. 81 as a function of the crack density.

In order to verify whether the choice of rather thick test layers has influenced the degradation the over-all results of the GFRP-test series were compared in Fig. 82 to the data from [Highsmith and Reifsnider 1982] measured with comparatively thin test layers (1.2 mm thick). This comparison shows perfect matching between the data sets.

In addition to the flat specimens tubular specimens were used to determine both the degradation of the Young's Modulus $E_{\perp}$ and the in-plane shear modulus $G_{\perp\parallel}$. The basic laminate lay up with supporting layers and an embedded test layer is similar to that of the flat specimens. A 89.45°-test-layer with a thickness of 3 mm is embedded in two ±45°-supporting layers (balanced angle ply) of 0.5 mm thickness (Fig. 83). These specimens are loaded in axial-direction (0°) until the first IFF occurs and unloaded afterwards. Subsequently they are subjected to a pure torsional moment without introducing further damage.

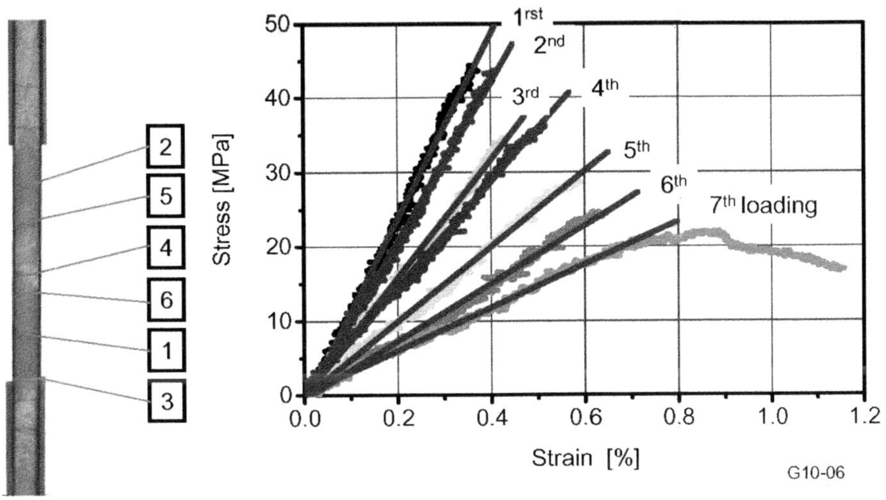

Fig. 81. Stress/strain curves of flat specimen

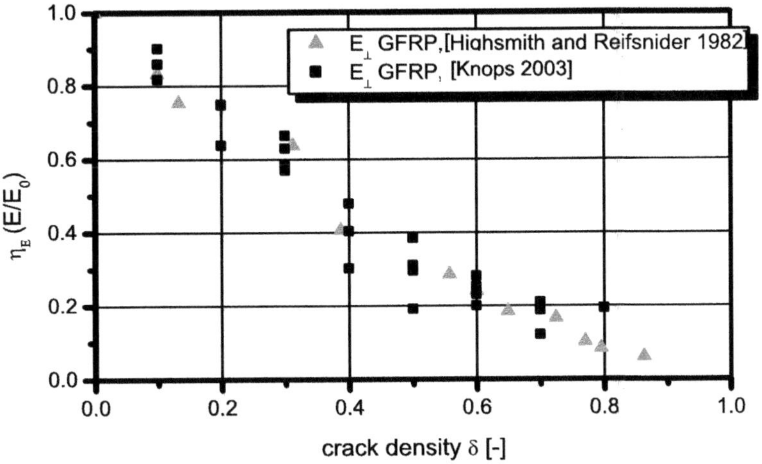

Fig. 82. Comparison of flat specimen results with [Highsmith and Reifsnider 1982]

In the same way the specimen is subjected to combinations of tension and torsion as well as compression/torsion, again without introducing further damage. From the recorded load/deformation data the stress/strain-diagrams of the test layer and thus $E_{\perp}$ and $G_{\perp\parallel}$ can be derived.

After the described loop the specimen is again loaded with pure axial tension until the next IFF occurs in the test layer and the procedure described above starts again. This is repeated until crack saturation. The

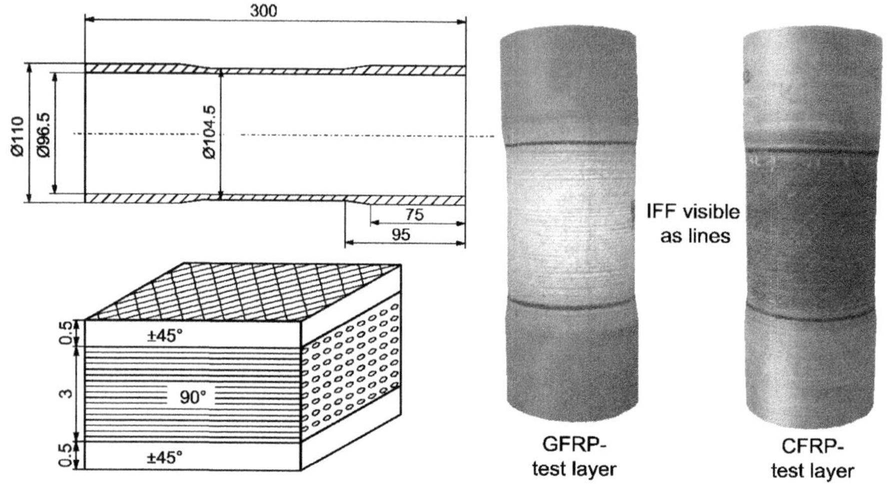

Fig. 83. Tubular specimens

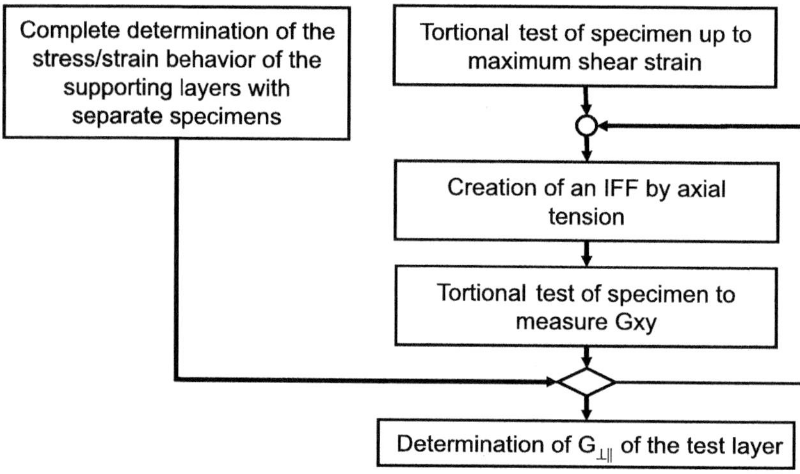

Fig. 84. Testing procedure to determine the degradation of $G_{\perp\|}$

complete procedure is illustrated in Fig. 84. In Fig. 83 the IFF in the test layer can be seen as circumferential lines through the translucent ±45° supporting layers.

Just as in the flat specimens IFF was detected optically, acoustically and by the sudden drop of stiffness. The optical detection is even possible for CFRP-test layers. Here the small delaminations caused by the IFF at the interface between CFRP test-layer und supporting GFRP-layer can be detected as thin lines surrounding the specimen (compare Fig. 83).

With the described procedure the IFF is in general not caused on the same load path the mechanical properties ($E_\perp$ and especially $G_{\perp\parallel}$) are measured on. Instead IFF is always caused by pure axial tension.

Nevertheless, this procedure is correct, because shear stressing $\tau_{\perp\parallel}$ and transverse tension $\sigma_\perp^t$ lead to the same fracture angle and thus the same damage (IFF Mode A and B). Moreover, the specimen behaves fully elastically for all loadings and unloadings as long as no further (micro-) damage is caused by the torsional loading [Li et al. 2000]. The degradation of the mechanical properties due to IFF is primarily depending on the crack density. It is of minor importance what stressing caused the IFF, as long as the fracture angle is correct (0° in the case examined here).

The degradation curves resulting from these tests for $E_\perp$ and $G_{\perp\parallel}$ are illustrated in Fig. 85 in dependence of the crack density. Obviously, the shear modulus is much less affected by IFF than the tensile Young's Modulus. This can be explained by friction on the crack surfaces. These results disprove Puck's original assumption saying that IFF has the same effect on both $E_\perp$ and $G_{\perp\parallel}$ [Puck 1996]. The observed degradation of $G_{\perp\parallel}$ can be also be predicted by micro-mechanics. In principle the experimental results confirm the calculations published for instance in [Hashin 1985, Li et al. 1998, Whitney 2000]. However, the micro-mechanical results seem to be an upper boundary and still overestimate the degradation of the shear modulus $G_{\perp\parallel}$.

Another point worth to mention is that the results for the Young's modulus obtained with the tubular specimens (Fig. 85) match perfectly with those deduced from the flat specimen tests. Generally, the experimental data illustrated in Fig. 85 prove that the ply discount method which degrades $E_\perp$ and $G_{\perp\parallel}$ to a fixed value after first IFF is very unrealistic.

In Fig. 61 the same experimental results are shown in a different way. Here, the stress exposure for IFF $f_{E,\,IFF}$ is used as x-coordinate. This is the decisive step for the calibration of phenomenological models as proposed by Puck.

Figure 61 can be derived straight forward from Fig. 85. All IFF in the specimens develop under pure transverse tension. In this simple case the stress exposure is calculated by the trivial equation $f_{E,\,IFF} = \sigma_2/R_\perp^t$. Here, σ_2 is the transverse tensile stress in the test layer and $R_\perp^t$ is the transverse tensile strength of the test layer. After IFF-initiation the – now "fictive" – stress σ_2 is calculated from the measured strains and the Young's modulus of the lamina at IFF-initiation. Thus, in the case of uniaxial tensile stressing a stress exposure of "2" means that the strain ε_2 in the test layer is

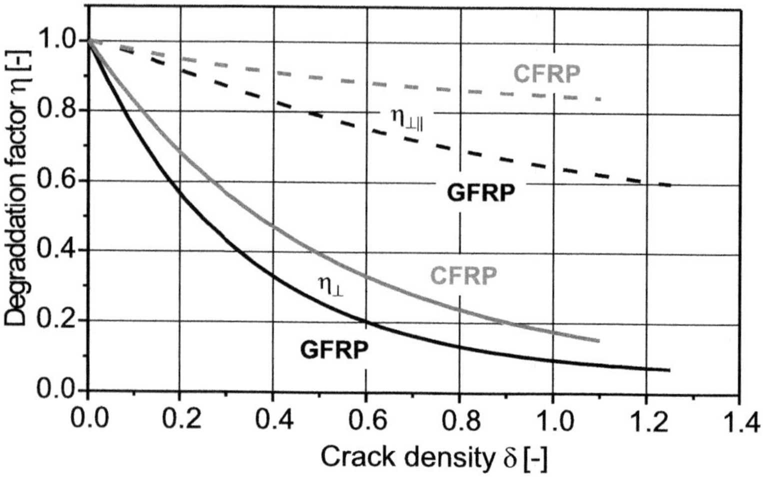

Fig. 85. Degradation curves as a function of the crack density [Knops Bögle 2006]

twice that at IFF-initiation. In general, after IFF-initiation $f_{E,\,IFF}$ can be calculated from the measured strains. In addition, in all experiments the crack density was recorded as a function of the measured strains. In combination this gives a straight forward correlation between crack density and $f_{E,\,IFF}$. With this correlation Fig. 61 can be derived from Fig. 85.

With the experimental data described here, the degradation parameters of Table 4 have been determined.

6.2.2 Degradation of $E_\perp$ and $G_{\perp\parallel}$ for fracture Mode B

Under transverse compressive stressing $\sigma_\perp^c$ the Young's modulus is – according to the common model – not degraded by IFF, because the fracture surfaces are pressed on each other. The shear modulus $G_{\perp\parallel}$ is less degraded in the regime of Mode B (($\sigma_\perp^c$, $\tau_{\perp\parallel}$)-combinations) because the superimposed compressive stress increases the friction between the fracture surfaces and more shear can be transmitted. With the tubular specimens (compare Fig. 83) the shear modulus $G_{\perp\parallel}$ has been determined for a series of different ratios $\sigma_\perp/\tau_{\perp\parallel}$. These experimental results (shown in Fig. 86 for GFRP; behavior of CFRP qualitatively the same) are in line with the explained model. The higher the normal stressing $\sigma_\perp^c$ superimposed to the shear stressing $\tau_{\perp\parallel}$ the higher the shear modulus (at constant crack density).

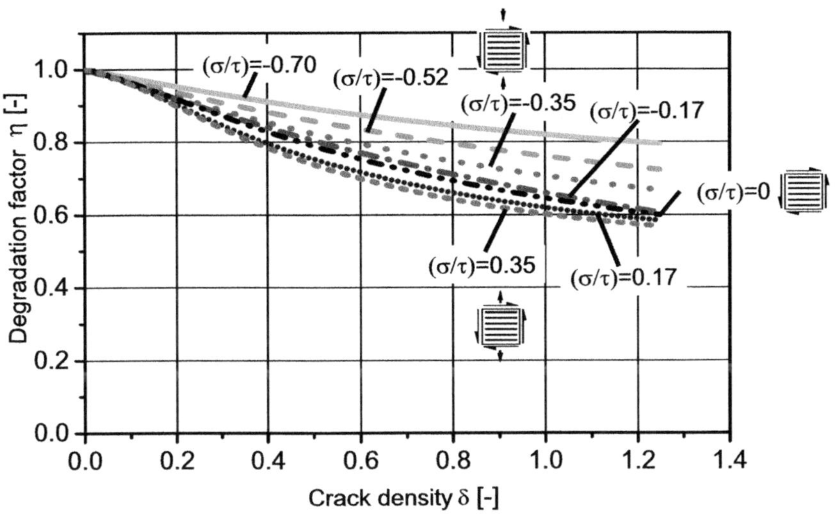

Fig. 86. Degradation curves $G_{\perp\parallel}$ (GFRP) for all stress ratios $\sigma_{\perp}/\tau_{\perp\parallel}$ [Knops 2003]

The Young's modulus $E_{\perp}^{c}$ is only independent of the crack density if the cracks are closed. However, in a multiaxial laminate there are tensile thermal stresses transverse to the fibers[23]. As a consequence IFF-cracks stay open when no mechanical loads are applied. Under compressive load the cracks have to be closed before compression can be transmitted. As long as the cracks are open $E_{\perp}^{c}$ is practically zero. This can clearly be seen in the stress/strain-curves measured for different crack densities (shown in Fig. 87 for GFRP). The higher the crack density the more the laminate can be compressed before the cracked layer takes over load and contributes to the stiffness of the laminate. In CFRP the cracks open wider, but the qualitative behavior is the same.

6.2.3 Degradation of $\nu_{\perp\parallel}$ and $\nu_{\parallel\perp}$

The Poisson ratios of a UD lamina ($\nu_{\perp\parallel}$ and $\nu_{\parallel\perp}$) are linked by (Eq. 123). Here $\nu_{\perp\parallel}$ described the contraction in "⊥" direction due to a stressing in "||" direction (definition used in [VDI 2006]). Because of this direct correlation between the two ratios it is sufficient to investigate the degradation of one

[23] At temperatures below T_G

of them. The larger ratio $\nu_{\perp\parallel}$ was chosen because of the higher – and therefore easier to measure – strains.

$$E_{\perp} \cdot \nu_{\perp\parallel} = E_{\parallel} \cdot \nu_{\parallel\perp} \qquad \text{(Eq. 123)}$$

Filament wound tubes with a 90° test layer were chosen as specimens. The specimens are loaded by internal pressure[24] and friction between fixture and specimen is minimized in axial direction in order to enable a free movement (in axial direction). Under these conditions, the 90° layer is – as desired – under pure σ_1 stress. The laminate consists – in addition to the test layer – of supporting layers with ±28° direction (with reference to the axial direction). These layers remain undamaged up to an axial strain of 1% and the Poison ratio ν_{xy}[25] matches that of the – undamaged – test layer.

Before the internal pressure tests the specimens are – in a separate test – subjected to axial tension in order to create IFF in the 90° layers. Under internal pressure the axial and circumferential strains are then measured (by several strain gauges) for the undamaged specimens and for 2 crack densities. Table 7 shows the test results for the Poisson's ratio of the laminate (with standard deviation). In addition the theoretical values for ν_{xy} are listed for the two postulates under investigation:

1. The "large" Poisson ratio $\nu_{\perp\parallel}$ is not influenced by IFF. In this case ν_{xy} is constant and equal to $\nu_{\perp\parallel}$ (because the supporting layers were designed to have $\nu_{xy} = \nu_{\perp\parallel}$)
2. The "large" Poisson ratio $\nu_{\perp\parallel}$ is degraded as the Young's modulus $E_{\perp}$ (parameters according to Table 4). In this case ν_{xy} gets smaller with increasing crack density.

Table 7. Poisson's ratio νxy of the [±28°, 90°]S-GFRP-laminate

Crack density δ [–]	ν_{xy} measured [–]	ν_{xy} calculated with $\nu_{\perp\parallel}$ = const. [–]	ν_{xy} calculated with $\eta_\nu = \eta_E$ [–]
0	0.244 (±0.014)	0.27	0.27
0.5	0.256 (±0.022)	0.27	0.22
1	0.267 (±0.003)	0.27	0.23

24 Oil is used as pressure medium. A silicon tube is used for sealing.
25 (global coordinate system, x = axial direction of the tubular specimen)

The results summarized in Table 7 show that the measured values for the undamaged specimens are slightly smaller than expected which might be due to the fact that friction could only be minimized, but not excluded. Thus the strain in axial direction is slightly smaller than in theory. This effect gets less relevant with increasing crack density, because the laminate in the test region is then less influenced by the interaction with the fixtures[26]. The measured ν_{xy} is for higher crack densities slightly higher than for the undamaged specimens and very close to the theoretical value calculated with the assumption that $\nu_{\perp\parallel}$ does not change with crack density.

Thus, according to the experimental data it seems reasonable not to degrade $\nu_{\perp\parallel}$. At this point it should be mentioned, that – in all cases which have been investigated by the author – the question of whether degrading $\nu_{\perp\parallel}$ or not has very small influence on lamina stresses and strains. Of superior importance is in any case the correct degradation of $E_\perp$ and $G_{\perp\parallel}$.

6.2.4 Validation of degradation curves

6.2.4.1 Experimental validation

The degradation parameters of Table 4 and the proper functioning of the software SUBU written in the context of the PhD-work documented in [Knops 2003] have been validated by a further series of experiments on tubular specimens under internal pressure.

Such fiber reinforced tubes loaded by internal pressure are more and more used for instance in the offshore industry. The tubes are often filament wound with ±55° (exactly: ±54,74°) fiber direction which is the correct angle according to netting theory. This orientation guarantees that the fibers alone can take over the stresses resulting from internal pressure. Because of this fact it is widely believed that the deformation behavior of these tubes is dominated by the fibers and thus linear elastic. This implies that possible IFF should not influence the deformation considerably.

However, just the opposite is the case. Figure 88 shows the stress/strain-curves of a GFRP tube with ±54.74°-fiber direction as it has been published by the organizers of the "World Wide Failure Exercise" [Soden et al. 2002] and as it has been calculated in the Failure Exercise[27]. The measured

---

[26] This is only an attempt to explain the observed behavior. The idea is that the cracks close to the fixtures, outside the test region, can open and compensate the friction in the region of the fixtures. This "theory" was not validated.

[27] The circumferential stress is proportional to the internal pressure. This form of presentation was chosen following the corresponding diagram in the Failure Exercise.

curves are far from linear and there is a clear disagreement between experimental data and numerical prediction. Obviously the deformation behavior of the tubes is dominated by non-linear effects which have not been

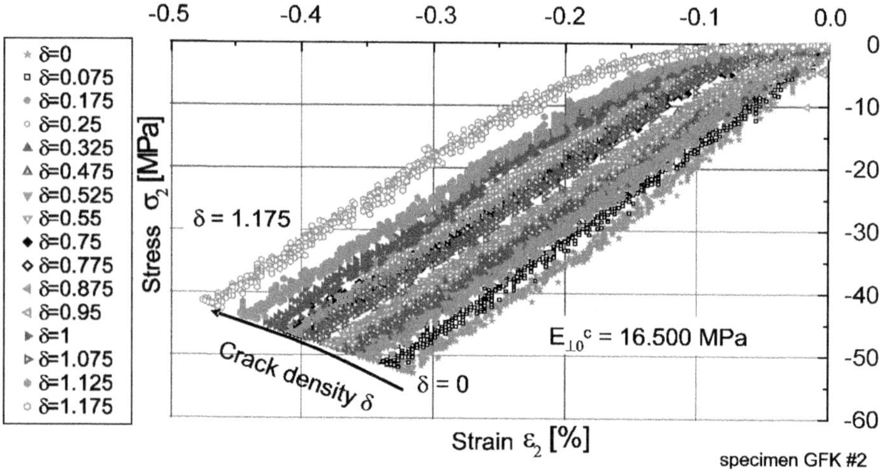

Fig. 87. Stress/strain-curves (σ_2^c, ε_2) (GFRP) for different crack densities [Knops 2003]

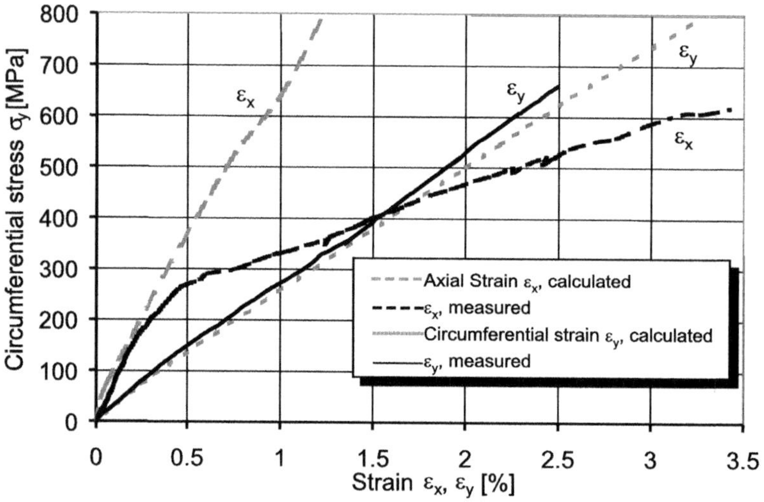

Fig. 88. ±54.75°-GFRP-tube under internal pressure: test data [Soden et al. 2002] and calculation [Puck and Schürmann 2002]

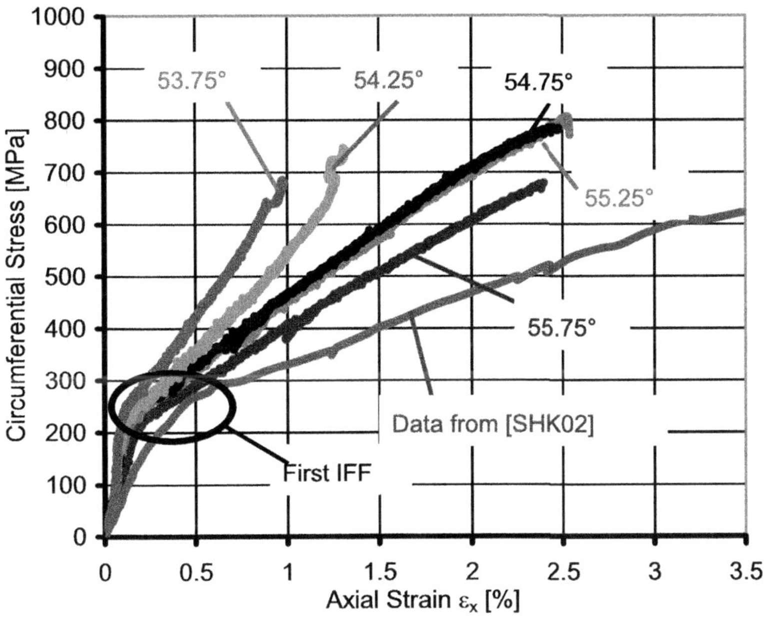

Fig. 89. Axial strain ε_x for different winding angles, GFRP

correctly covered by any of the models applied in the failure exercise [Hinton et al. 2002]. The non linear effects are obviously due to IFF. However, the correctness of the experimental data itself was queried in [Soden et al. 2002], too.

These considerations show that tubes under internal pressure are well suited to verify whether the degradation curves presented above do help calculating the deformation of FRP-structures more accurately.

Moreover, it was at that time a goal to verify the experimental data supplied in [Soden et al. 2002] for the World Wide Failure Exercise. In the context of the investigations specimens similar to those used for the failure exercise were fabricated with different fiber directions varying between ±53.75° and ±55.75°. The thickness of the layers is 0.25 mm each. Figure 89 shows the axial strain measured for the different winding angles in a diagram similar to Fig. 88. Clearly the point of IFF-initiation can be detected from the curves. However, the degree of non-linearity and especially the effect of small changes in fiber direction are astonishing. Obviously the specimens tested for the Failure Exercise [Soden et al. 2002] have been wound with angles above ±55° (probably ±56°).

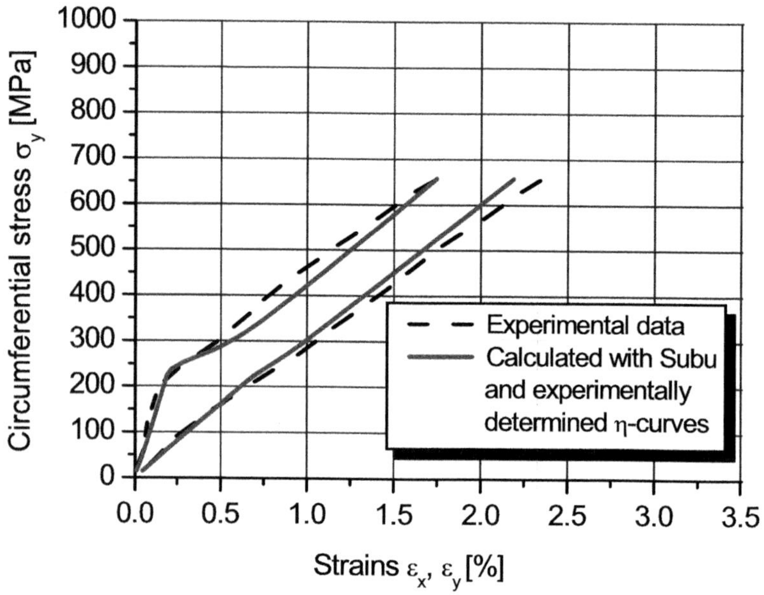

Fig. 90. Axial and circumferential strains calculated with SuBu versus experimental data for ±54.75°, GFRP

Figure 90 shows the summarized experimental results for the fiber direction of ±54.75° and stress/strain-curves which have been calculated with *Subu*. There is an almost perfect matching between experiment and simulation. The degradation parameters for the analysis were taken from Table 4. The material data comes from [Soden et al. 1998][28]. Only the transverse tensile strength $R_{\perp}^{t}$ and the longitudinal shear strength $R_{\perp\parallel}$[29] were not taken from [Soden et al. 1998] but instead directly determined from the internal pressure tests (compare Table 8).

Obviously, the limit of IFF-initiation is correctly calculated in Fig. 90 because it depends on the ratio of $R_{\perp}^{(t)}$ and $R_{\perp\parallel}$ and not as much on their magnitude. The strength values were taken from the test results. However, the correct calculation of the deformation after IFF-initiation is unique and shows that the degradation of stiffness due to continuous IFF-formation is modeled correctly.

[28] data for UD-lamina Silenka E-Glass/Epoxy MY750 / HY917 / DY063

[29] $R_{\perp}^{(t)} = 74$ N/mm^2, $R_{\perp\parallel} = 105$ N/mm^2

Table 8. Comparison of transverse tensile strength R⊥(t)

	Layer-thickness [mm]	$R_\perp^{(t)}$ [N/mm^2]	No of spec. [–]	Standard deviation [N/mm^2]	$R_\perp^{(t)}{}_{Lam}$/ $R_\perp^{(t)}{}_{SL}$ [–]
GFRP					
Flat specimens, single layer (SL)	5	40.5	5	4.2	
Flat specimens, embedded	5	50.5	8	12.6	1.25
Flat specimens, SL	10	32.5	10	7.1	
Flat specimens, embedded	10	40.0	7	4.6	1.23
Tubular specimens (compare Fig. 83)	3	46.0	4	7.4	
Tubes from internal pressure tests	0.25	≈74	7	1	
CFRP					
Flat specimens, single layer (SL)	5	18.1	5	5.8	
Flat specimens, embedded	5	26.6	5	13.1	1.47
Tubular specimens (compare Fig. 83)	3	41.6	4	1.4	
Tubes from internal pressure tests	0.25	≈64	7	0.8	

6.2.4.2 Numerical validation by means of FEM

The experimental results on the degradation of $E_\perp$, $G_{\perp\parallel}$ and $\nu_{\perp\parallel}$[30] go in line with fracture and damage mechanical calculations (compare [Li et al. 1994]). A way to numerically verify the degradation of stiffness is a detailed FEM-analysis. Lambrecht [Lambrecht 2007] conducted such an analysis and compared the numerical results to the experimental work of Knops [Knops 2003]. Figure 91 illustrates the FEM-model used for the calculations. The used mesh of solid elements (ANSYS® Solid 186) is very fine.

Cracks are modeled by attaching the nodes of the crack surface only to the elements on one side of the surface. Different variants of loads, laminates (characterized by the properties of the tree layers and the relative layer thicknesses) and crack density (varied by the distance between the

[30] $\nu_{\perp\parallel}$ is not degraded by IFF

Table 9. Material properties used for the supporting layers (0° and ±45°) in the FEM-analysis; coordinates according to Fig. 91 [Lambrecht 2007]

	Property	Fiber direction $\alpha = 0°$	Fiber direction $\alpha° = °\pm 45°$
Young's modulus $[N/mm^2]$	E_x	45.600 ($= E_{\parallel}$)	17.663
	E_y	16.200 ($= E_{\perp}$)	17.663
	E_z	16.200 ($= E_{\perp}$)	16.200
Shear modulus G $[N/mm^2]$	G_{xy}	5830 ($= G_{\perp\parallel}$)	13.571
	G_{yz}	5786 ($= G_{\perp\perp}$)	5786
	G_{xz}	5830 ($= G_{\perp\parallel}$)	5830
Poisson's ratios	ν_{xy}	0.278 ($=\nu_{\perp\parallel}$)	0.5148
	ν_{yz}	0.4	0.4
	ν_{xz}	0.278 ($=\nu_{\perp\parallel}$)	0.278

cracks) were investigated in the study. The mechanical properties used for the supporting layers are listed in Table 9.

All results presented here have been calculated using material properties of GFRP for all layers. Figure 92 illustrates the degradation of the Young's modulus ($\eta_E = E_{\perp}/E_{\perp}^0$) for different thicknesses of the 90°-layer (which is damaged by IFF). There is only a minor influence of the layer-thickness on the calculated curve. This supports the concept of using specimens with relatively thick test-layers in the experimental program.

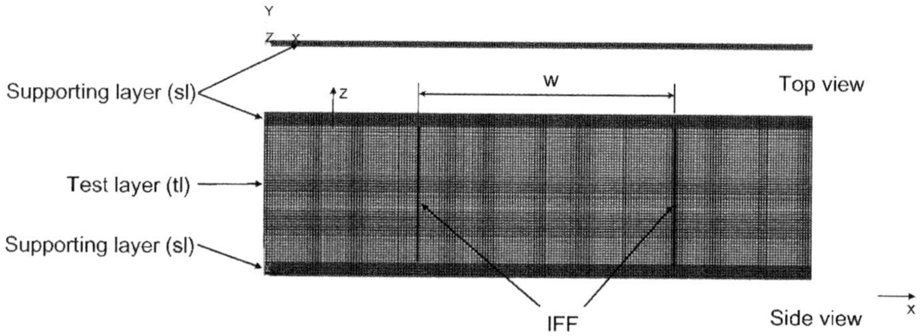

Fig. 91. FEM-model for the numerical validation of the stiffness degradation [Lambrecht 2007]

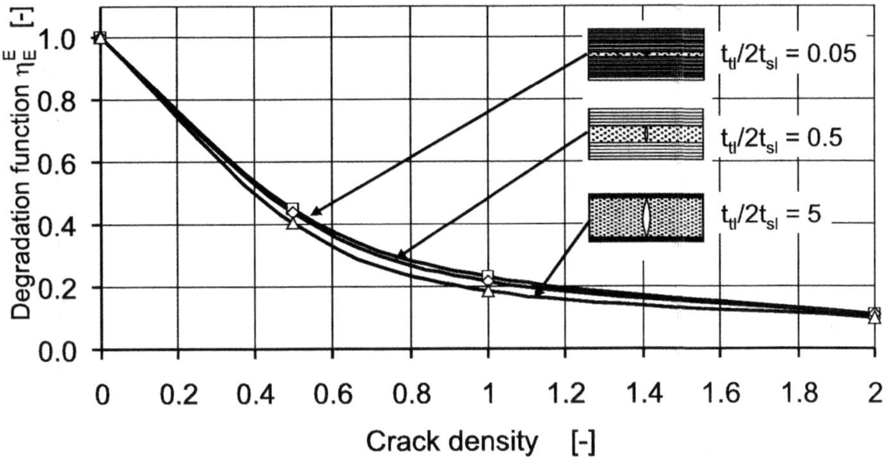

Fig. 92. Degradation curves modeled by FEM for different thicknesses of the 90° layer [Lambrecht 2007]

Figure 93 shows the result of another parametric study. Here the stiffness of the supporting layers was varied and the influence of the supporting layers on the degradation of $E_\perp^t$ studied. There is almost no measurable influence of the laminate lay-up on the degradation of $E_\perp^t$ as long as the cracked layer is embedded in the laminate.

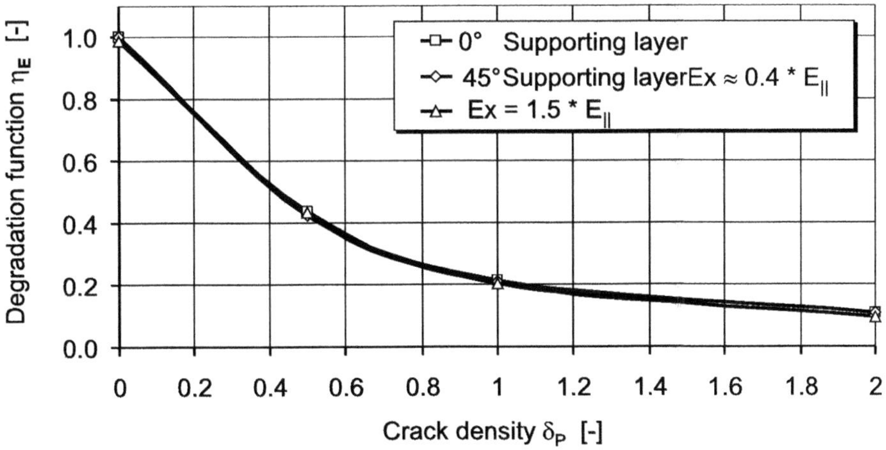

Fig. 93. Degradation curves modeled by FEM for different laminate lay-ups [Lambrecht 2007]

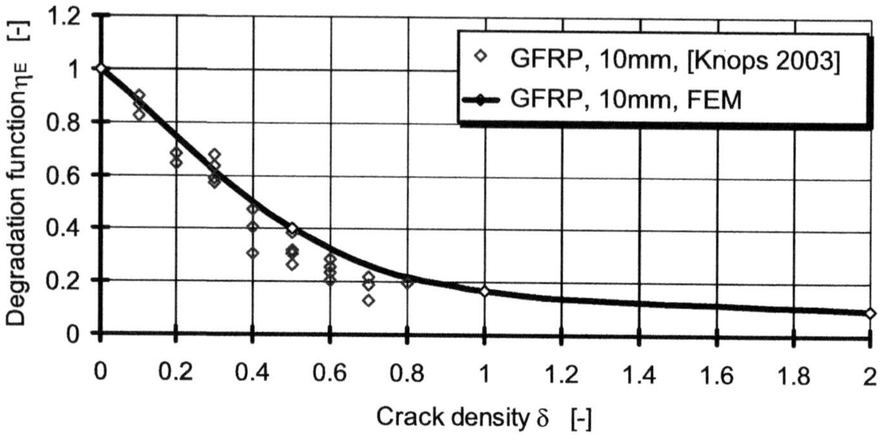

Fig. 94. Comparison of experimental results and FEM-results on ηE [Lambrecht 2007]

Figure 94 finally compares the numerical results with the experimental data. The FEM-results are – for the same laminate lay-up – a lower boundary to the experimental results, meaning that – for higher crack densities $\delta > 0{,}3$ – the measured Young's modulus $E_{\perp}^{t}$ is a bit lower than the calculated one. This can be explained by the fact, that in reality the stiffness is not only reduced by the crack itself (as assumed in the FEM-model). Additionally, there are small delaminations at the tip of the crack which add to the reduction in stiffness, too.

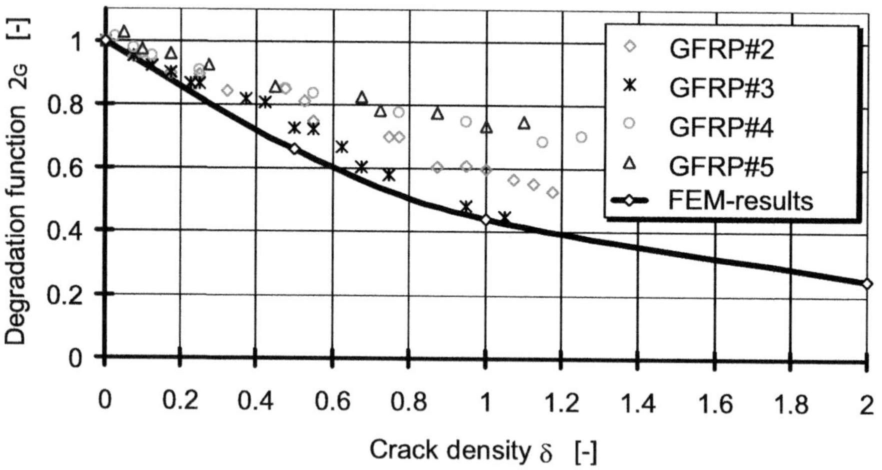

Fig. 95. Comparison of experimental results and FEM-results on η_G [Lambrecht 2007]

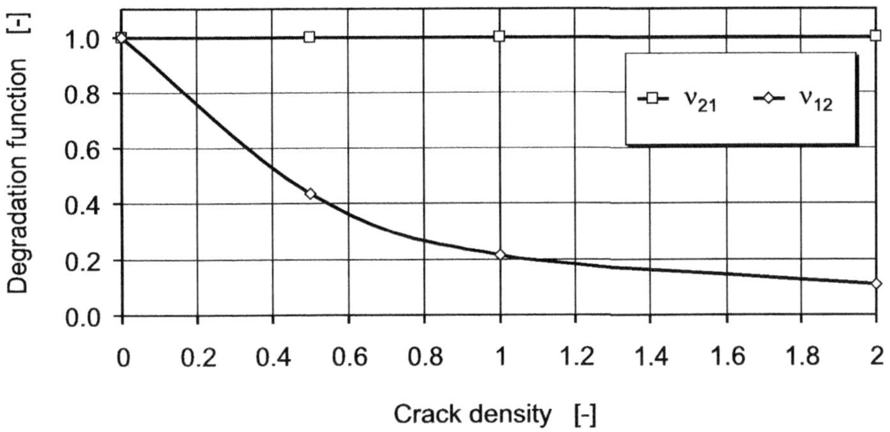

Fig. 96. Comparison of experimental results and FEM-results on the Poisson's ratio $\nu_{\perp\|}$ [Lambrecht 2007]

For the shear modulus $G_{\perp\|}$ the comparison of the numerical results with the experimental data is given in Fig. 95. The FEM-results, too, show that the shear modulus $G_{\perp\|}$ is much less affected by IFF than the Young's modulus $E_{\perp}^{t}$. However, the FEM-curve for η_G ($G_{\perp\|}/G_{\perp\|}^{0}$) form an upper boundary to the experimental results, meaning that the measured shear stiffness $G_{\perp\|}$ is a bit higher than the calculated one. This can be explained by the fact that the friction between the crack surfaces is not included in the FEM-model.

Another interesting result of the FEM-study concerns the Poisson's ratios. Here, too, the experimental results of Knops, suggesting that there is no degradation of the major Poisson's ratio $\nu_{\perp\|} = \nu_{21}$, are asserted by the study. Figure 96 shows the corresponding numerical results for the Poisson's ratios. Thus it can be concluded that there is no degradation of $\nu_{\perp\|}$.

6.3 Summary of chapter

The first part of the chapter on experimental work covers the validation of the Puck criteria. Huybrechts and Kopp [Cuntze et al. 1997, Huybrechts 1996, Kopp 2000] experimentally verified for the first time a fracture criterion for fiber reinforced composites both under 2D- and 3D-stressing. As a result not only the fracture hypotheses of Puck are validated. The correct determination both of fracture limits and fracture angles was verified, too. In more detail the evaluation of the experimental work is the following:

- Puck's IFF-criteria define fracture curves which are "open" for biaxial compressive stress $\sigma_\perp^c$ (compare Fig. 69). However, the UD-lamina cannot withstand unlimited compressive stress. Based on a micromechanical FEM-analysis and hydrostatic tests a damage threshold could be found. Above this threshold severe micro cracks develop which cannot be tolerated.
- The torsional ring tests showed that the fracture angle continuously increases from $\theta_{fp} = 45°$ for pure $\tau_{\perp\perp}$-stressing to $\theta_{fp} = 90°$ for pure $\tau_{\perp\parallel}$-stressing. This proves, that the fracture mode A*, where $(\sigma_n^t, \tau_{nt}, \tau_{n1})$-stress combinations on the fracture plane cause fracture, really exists for transverse isotropic materials. The fracture angles from experiment and theory match perfectly.
- The fracture curve of the Puck criteria matches perfectly with the fracture limits measured. This was extensively verified with tubular specimens in [Cuntze et al. 1997] for several $\sigma_\perp/\tau_{\perp\parallel}$-ratios.
- The postulated mechanical interaction of the shear τ_{nt} and τ_{n1} and the fracture impeding influence of σ_n^c could be validated experimentally with the slant shear specimens. With these experiments the fracture angles calculated by Puck for $(\tau_{\perp\parallel}, \tau_{\perp\perp})$-combinations could be validated, too.
- In addition to the slant shear tests the transverse compression tests proved that σ_n^c impedes shear fracture by $\tau_{n\psi}$ stress.

All together, the experimental work of Huybrechts and Kopp validates the fracture hypotheses of Puck as the correct physical basis of the fracture criteria in addition to the verification of the quantitative results both on fracture limits and fracture angle.

Great effort has not only been put into the validation of Puck's fracture criteria, but also on the determination of degradation curves for the elastic properties of embedded UD-layers being damaged by IFF. In the second part of the chapter this work is presented in detail. Both for GFRP and CFRP the degraded Young's modulus $E_\perp$ and shear modulus $G_{\perp\parallel}$ have been determined for all crack densities and corresponding stress exposure $f_{E,\,IFF}$. The major Poisson's ratio remains uneffected by IFF. The degradation curves are in good approximation valid for all laminate lay-ups independent of the position of the cracked layer within the laminate. The curves can also be used to verify or calibrate other degradation models.

The degradation curves are in general valid as long as the effected layer is not under transverse compression (as long as $\sigma_2 \geq 0$). Under compression, the crack surfaces are pressed on each other and the degradation of stiffness is less significant. For this case the degradation has been investigated, too.

The experimentally found degradation curves have been validated by additional experiments as well as by numerical studies. The results of this work are presented here.

7 Implementation in software

Talking about software for stress and fracture analysis of composite structures one has to distinguish between analytical tools and FEM-tools. Even though FEM-tools offer superior possibilities for a detailed analysis of complex structures, analytical tools are still valuable in engineering practice for pre-dimensioning of designs, but also – often in combination with FEM-tools – for design iterations, laminate optimization, the post processing of FEM-results and other design tasks which can be very time consuming if only FEM is used.

In general, both analytical and FEM-tools offer in addition to laminate stress/strain-analysis some kind of fracture analysis for composites. The standard fracture criterion used for this analysis is in most cases still the Tsai/Wu-criterion. However, in more and more commercial analytical tools like *ESAcomp*® (*www.componeering.com*) and *Composite Star*® *(www.material.be)* the Puck criterion (2D and 3D) is meanwhile implemented. Comparable freeware or semi-commercial tools with implemented Puck criteria are for instance the programs *AlfaLam* of the Technical University of Darmstadt (free download under *www.klub.tu-darmstadt.de*) and *Compositor* (for information: *www.ikv-aachen.de*). Both tools are based on Excel©.

Most FEM-tools have still only implemented the Tsai/Wu global criterion as standard. However, some activities are on the way and this situation might change soon. In the meantime, in many companies users have implemented the Puck criteria as a post processing tool to FEM programs. The implementation of the Puck 2D-criteria with subsequent calculation of the stress exposure or the stretch factor for the load determined stresses (see Chapter "Calculation of the stretch factor f_S^L of the load-determined stresses") is a rather easy task and the result including the fracture angle is obtained straight forward. The implementation of the 3D-criteria asks for some more programming and computational time, because the UD-stresses need to be transferred to the action-plane-coordinate system and for each knot/element the state of stress has to be analyzed for all potential fracture planes. This makes for each knot/element 180 calculations, if the analysis is done in 1°-steps. However, with today's computer technology this numerical effort is no longer a problem.

Very rare are so far tools which include a state of the art post failure analysis of the laminate based on the Puck theory in their routines. Most scientists working in this field have developed their own (scientific) tools, which allow the use of their own method. Examples for that are the software tools Li, Reid and Soden [Li et al. 2000] or Puck and Schürmann [Puck and Schürmann 1998, Puck and Schürmann 2002] used for the World-Wide Failure Exercise.

The FEM-System *STAGS* [Knight et al. 2002] has meanwhile integrated a progressive fracture process. The user can chose between several ply-discount models ($E_\perp$ and $G_{\perp\parallel}$ of a layer with IFF is set to zero) and a damage mechanical model [Chang and Chang 1987]. The accuracy of the computational results is right now limited by the quality of the implemented models, but the gradual failure process as such works correctly.

At the University of Stuttgart a research group is currently implementing Puck's complete theory not only including the criteria themselves, but also all the work presented in the chapter "Extensions to the IFF-criteria" and the degradation model in the FEM-software Abaqus. This tool will be made available to all Abaqus users. First results of this work are presented in [Busse at al. 2006]. Meanwhile a comprehensive paper of this work has been submitted for Part A of the World Wide Failure Exercise No2 [Deuschle and Kröplin 2008].

In the analytical tools *AlfaLam.nl* (free download under *www.klub.tu-darmstadt.de*) and *Compositor* the post failure analysis of the laminate according to Puck (compare chapter "Puck's approach for the analysis of the gradual failure process") has already been integrated.

The tool *AlfaLam.nl* replaces the software NOLI FRAN COLAM (non-linear fracture analysis of composite laminates) which was used by Puck and Schürmann in the World Wide Failure Exercise No1 (WWFE No1) [Puck and Schürmann 1998 and Puck and Schürmann 2002]. NOLI FRAN COLAM was written in FORTRAN IV, which is not common anymore. Mainly for this reason a more modern software became necessary. AlfaLam.nl has a modern user interface and is in general easy to use. But also on the scientific side there are some improvements. For instance, an elegant method of incorporating a weakening effect due to σ_1 as it is described in this book is implemented as well as the "Puck-Mannigel"-approach for strain interactions of combined (σ_2, τ_{21})-stresses (compare Annex and [Puck and Mannigel 2007]). Compared to NOLI FRAN COLAM there is also a difference in the methods of stiffness degradation after IFF-onset. In the WWFE No1 Puck and Schürmann degraded $E_\perp$ and

$G_{\perp\parallel}$ with a common factor by keeping the stress exposure $f_E = 1$ constant after IFF-onset. In *AlfaLam.nl* the experimental results of [Knops 2003] are implemented. Thus, $G_{\perp\parallel}$ is considerably less degraded than $E_{\perp}$ as it is described in [VDI 2006].

The software *AlfaLam.nl* can be downloaded from the homepage of the institute klub of the Technical University of Darmstadt (*http://www.klub.tu-darmstadt.de/forschung/download.php*). There is also a very detailed manual available on the homepage. The user can add own modules to the code which is Excel-based.

8 Application of Puck's work in industrial practice

The Puck criteria are meanwhile widely used and accepted as state of the art in IFF-analysis. A number of guidelines refer to the Puck criteria. This list includes the VDI-guideline VDI2014, Part 3 [VDI 2006] on the Development of FRP components, the Guideline for the Certification of Wind Turbines of Germanische Lloyd [Germanische Lloyd 2003] and the Det Norske Veritas Guideline DNV-RP-F202 on the use of Composite Risers in Offshore applications [DNV 2003].

In contrast to this, the application of the complete Puck theory including the post failure analysis is not yet as widely spread. For some standard problems in composite design such a detailed analysis is indeed not needed. However, cases for which only a state of the art post failure analysis would detect the realistic behavior of the structure are often not even detected as critical. A simple example of this is a $\pm 55°$ tube for internal pressure as it is discussed in the chapter "Experimental work". The use of design and FEM-software which has the complete theory implemented would for sure help avoiding failure of structures in practice which today happen in some cases because the structural problems have not been realized in the design phase, because they are only detectible by very experienced Composite designers and/or the application of a state of the art post failure analysis.

9 Concluding remarks

Puck's theory has evolutionarily developed over half a century starting with Puck's first work on the strength of GFRP-gliders, which was – just as the gliders themselves – revolutionary at the time. This was followed by a period of intensive scientific work from 1960 to 1970 at the "Deutsches Kunststoffinstitut" (DKI) in Darmstadt. This period cumulated in a really path breaking publication in 1969 [Puck 1969]. Puck was however during this decade continuously inspired by the growing experience resulting from fracture tests with wings and fuselages of modern GFRP-gliders.

The time at the DKI was followed by composite design work in industry, namely at Ciba Geigy where Puck worked especially on high voltage equipment, and Kassel University. Here Alfred Puck worked as Professor for Composite Technology especially on the development of highly loaded GFRP-components for automotive industry. A new advancement in the theory was announced in 1992 after Puck's retirement from his job as Professor at Kassel University by the paper "A fracture criterion shows the direction" [Puck 1992a] followed by Puck's book on the strength analysis of fiber-matrix composites from 1996 [Puck 1996]. Meanwhile, at the Institute of Plastics Processing (IKV) in Aachen six dissertations have been written experimentally validating Puck's theory to an extent which is unprecedented. Just recently, Puck added some "chapters" to his theory, namely the inclusion of stresses not acting on the fracture plane in the strength analysis, the sophisticated calculation of the stretch factor of the load-determined stresses in the presence of residual stresses and the consideration of the strain interactions for combined (σ_2, τ_{21})-stresses (compare [VDI 2006, Puck and Mannigel 2007]).

Without any doubt, the theory is ambitious, but it is essentially not complicated, but straight forward and understandable for all design engineers. In general, composites are not "a" material, instead each laminate is an individually designed material. Such complex materials ask for an adequate theory which realistically covers the interaction of fibers and matrix.

Having read this book, the reader will have found out that the Puck theory is in the form presented here very flexible and adaptable to experimental data and special material properties. The flexibility to be precisely adaptable to experimental data means on the other hand a lot of parameters. Take as examples the inclination parameters $p_{\perp\parallel}^{t}$, $p_{\perp\perp}^{t}$, $p_{\perp\parallel}^{c}$; $p_{\perp\perp}^{c}$; the degradation factor η_{w1} with the parameter "s" as starting point of the weakening due to σ_1 and the parameter "m" to characterize the maximum weakening or the consideration of the effect of microdamage, quantified by the parameter Δ_{max}. At first glance, this multiplicity of parameters adds to the first impression, that the Puck theory is complicated. However, with a very limited set of parameters which is not larger than for other theories, the Puck theory can be used. Moreover, for all parameters, reasonable values are given here. In other words, the user does not have to determine all parameters himself, but can use default values.

As a conclusion I want to quote Alfred Puck saying that there is for sure no deficit of good theories, but instead a very obvious deficit of reliable experimental results. To overcome this still remaining deficit industry and scientific laboratories should work together all over the globe.

Annex

Nonlinear stress analysis before IFF

The $\tau_{21}(\gamma_{21})$ curve as well as the $\sigma_2(\varepsilon_2)$ curve for the compression domain of GFRP and CFRP show a distinct nonlinearity. The main reason for this is microdamage within the matrix and at the fiber/matrix interface which develops above a certain stress level but long before the macroscopic IFF. In a realistic lamina-by-lamina fracture analysis the nonlinearity due to τ_{21} and σ_2 compressive stress has to be taken into account. A prerequisite for this task is the availability of the $\tau_{21}(\gamma_{21})$ diagram for pure τ_{21} shear and of the $\sigma_2(\varepsilon_2)$ diagram for uniaxial σ_2 compressive stress.

In the course of a non linear lamina-by-lamina fracture analysis the loads acting onto the laminate are usually applied in small load steps. Instead of using the engineering *constants* Young's modulus $E_\perp$ and longitudinal shear modulus $G_{\perp\parallel}$ one makes use in the nonlinear fracture analysis of the secant moduli $E_{\perp s}$ and $G_{\perp\parallel s}$ which are taken from the corresponding stress-strain diagrams. The stress and strain analysis has to be repeated (iterated) for the load step under consideration until the calculated stresses and strains do not change significantly any more. Within each step of iteration a new value for $E_{\perp s}$ and $G_{\perp\parallel s}$ has to be taken from the corresponding stress-strain curve.

As a first result of the laminate analysis one obtains the strains ε_x, ε_y, γ_{xy} and the corresponding strains ε_1, ε_2, γ_{21} of each lamina. Based on these strains the lamina stresses are obtained by means of the Classical Laminate Theory (CLT). Here, the shear stress τ_{21} depends solely on the shear strain γ_{21} and is not related to ε_1 and ε_2. Therefore it is reasonable to explain some fundamental aspects with the help of the $\tau_{21}(\gamma_{21})$ curve. Normally one uses the initial (tangent) moduli $E_{\perp k}$ and $G_{\perp\parallel k}$ for the first calculation run of a laminate analysis. Figure 97a) shows the known $\tau_{21}(\gamma_{21})$ diagram of a lamina. Also shown are the lamina results of the first calculation run of a laminate analysis, namely the strain $\gamma_{21}^{(1)}$ and the stress $\tau_{21}^{(1)}$.

There are two different ways to determine the secant modulus $G_{\perp\parallel s}$, which shall be used for the following iteration step.

The shear strain $\gamma_{21}^{(1A)}$ is taken from the $\tau_{21}(\gamma_{21})$ curve at the calculated shear stress $\tau_{21}^{(1)}$ and the secant modulus $G_{\perp\|s}$ is calculated from $G_{\perp\|s} = \tau_{21}^{(1)}/\gamma_{21}^{(1A)}$, Fig. 97a).

The shear stress $\tau_{21}^{(1B)}$ is taken from the $\tau_{21}(\gamma_{21})$ curve at the calculated shear strain $\gamma_{21}^{(1)}$ and the secant modulus $G_{\perp\|s}$ is calculated from $G_{\perp\|s} = \tau_{21}^{(1B)}/\gamma_{21}^{(1)}$, see Fig. 97a).

The procedure B) is the preferred one. The convergence is faster because laminate strains (in contrast to stresses) do (in a multi-layer laminate) not change significantly when a modification of $E_{\perp s}$ and $G_{\perp\|s}$ takes place. Besides this there may be the risk with procedure A) that one fails to capture the stress-strain curve when trying to pick γ_{21} at a relatively high stress τ_{21}.

In case of σ_2 the stress analysis is more complicated. Because of the Poisson effects σ_2 as well as σ_1 are related to both strains ε_2 and ε_1.

From this one obtains the following relations:

$$\sigma_2 = E_{\perp s}(\sigma_2) \cdot \varepsilon_2 + \nu_{\|\perp}\sigma_1 \qquad \text{(Eq. 124)}$$

or $$\sigma_2 = E_{\perp s}(\sigma_2) \cdot \frac{\varepsilon_2 + \nu_{\perp\|}\varepsilon_1}{1 - \nu_{\perp\|}\nu_{\|\perp}} \qquad \text{(Eq. 125)}$$

or abbreviated $$\sigma_2 = E_{\perp s}\left(\sigma_2\right) \cdot \varepsilon_{2,1}^{comb} \qquad \text{(Eq. 126)}$$

For this reason it is not allowed to determine the secant Modulus $E_{\perp s}(\sigma_2)$ from a $\sigma_2(\varepsilon_2)$ curve resulting from a uniaxial test at that strain ε_2, which is calculated for combined (σ_1, σ_2) stress. Doing this the influence of σ_1 on σ_2 would be neglected, see (Eq. 124) and Fig. 97b).

On hand of the abbreviated formulation by introducing the so called "combined strain" $\varepsilon_{2,1}^{comb}$ in (Eq. 126) it becomes obvious that the stress σ_2 occurring simultaneously with σ_1 can formally be calculated as for a uniaxial σ_2 stress state by means of Hooke's law $\sigma = E \cdot \varepsilon$, however using $\varepsilon_{2,1}^{comb}$ instead of ε_2. If a uniaxial σ_2 stress state would result in a strain of the magnitude $\varepsilon_{2,1}^{comb}$, the stress σ_2 would correspond to the one which is reached in the laminate analysis for the combination of ε_2 and ε_1.

In most cases the stress-strain curve gained from a uniaxial test is expressed analytically by a relationship $\sigma_2 = F_\sigma(\varepsilon_2)$. In case of the biaxial (σ_1, σ_2) stress state the stress σ_2 can be calculated from $\sigma_2 = F_\sigma(\varepsilon_{2,1}^{comb})$. Thus the secant modulus $E_{\perp s}$ follows from $E_{\perp s(\sigma_2)} = \sigma_2/\varepsilon_{2,1}^{comb}$. This procedure is based on the assumption, that the modulus reduction from $E_\perp$ to $E_{\perp s}$

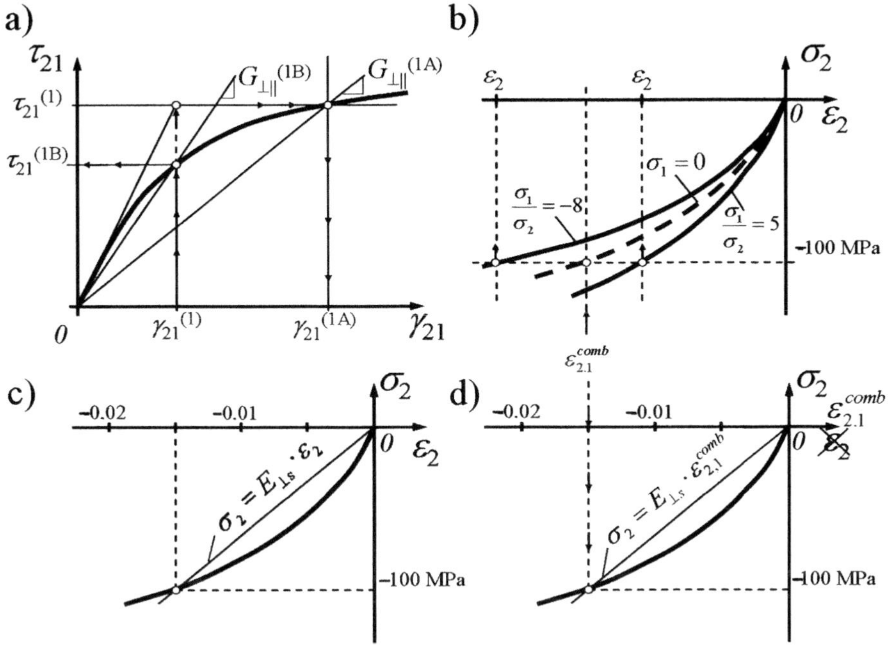

Fig. 97. Non-linear stress/strain-curves: a) $\tau_{21}(\gamma_{21})$ shear stress-strain curve with picking the secant modulus A) at the calculated stress $\tau_{21}^{(1)}$; B) at the calculated strain $\gamma_{21}^{(1)}$ b) $\sigma_2(\varepsilon_2)$ diagram for combined (σ_1, σ_2) stress. Equal stress σ_2 is reached at equal magnitude of $\varepsilon_{2,1}^{comb}$, but – depending on the ratio σ_1/σ_2 – at different strains ε_2. c) For the stress calculation only a $\sigma_2(\varepsilon_2)$ curve is available which is gained from a test with uniaxial σ_2 stress. d) For the stress calculation at combined (σ_1, σ_2) stress the curve from the uniaxial test can be used without any modification of the strain axis, however the values on the strain axis have changed their meaning from ε_2 to $\varepsilon_{2,1}^{comb}$.

depends solely on the stress σ_2, irrespective of how the stress σ_2 is achieved, whether e. g. by a uniaxial or a biaxial state of stress.

In order to determine $E_{\perp s}$ for the next iteration step one makes also use of the procedure B) which was recognized on hand of $G_{\perp\|s}$ as being favorable. More generalized this procedure can be described as follows:

If in the iteration step i the strains $\varepsilon_{1k}^{(i)}$ and $\varepsilon_{2k}^{(i)}$ as well as the shear strain $\gamma_{21k}^{(i)}$ were calculated for a lamina k, then the secant moduli $E_{\perp sk}^{(i+1)}$ and $G_{\perp\|s}^{(i+1)}$ for the iteration i+1 are determined by means of $\varepsilon_{2,1k}^{comb}$ in the following way: From the $\sigma_2(\varepsilon_2)$ curve or from the $\tau_{21}(\gamma_{21})$ curve one picks the stresses belonging to $\varepsilon_{2.1k}^{comb\ (i)}$ or $\gamma_{21k}^{(i)}$ respectively and calculates $E_{\perp sk}^{(i+1)}$ und $G_{\perp\|sk}^{(i+1)}$ from the already picked stresses and $\varepsilon_{2.1k}^{comb\ (i)}$ or $\gamma_{21k}^{(i)}$ respectively.

Analytical treatment of strain-interactions of combined σ_2 and τ_{21}

In reality the correlations between microdamage and secant moduli lead to more complicated relations than those shown above. Deformation measurements on laminae under (σ_2, τ_{21}) stress combinations revealed that σ_2 has also an influence on the $\tau_{21}(\gamma_{21})$ curve and that on the other hand τ_{21} does influence the $\sigma_2(\varepsilon_2)$ curve [Puck 1996, Kopp et al. 1997, Kopp 2000]. In a more recent paper a computational approach has been presented to take these effects into account by means of a simple physically based model [Puck and Mannigel 2007]. The fundamental idea of this model is that not only the dominating stress σ_2 or τ_{21} is relevant for the secant modulus $E_{\perp s}$ and $G_{\perp\parallel s}$ respectively, but that also the other stress which is present in a combined stress state affects the stress-strain curve due to its contribution to the microdamage in the composite. Based on this idea the secant moduli $E_{\perp s}$ and $G_{\perp\parallel s}$ are directly calculated from a mathematical function of the IFF-stress exposure $f_{E,IFF}$, see (Eq. 71), (Eq. 73) and (Eq. 77), which is caused by the combined action of σ_2, τ_{21}, and σ_1.

Microdamage caused by σ_2 alone is expressed by the stress exposure $f_E^{(\sigma_2)}$, calculated with σ_2 alone. Microdamage caused by τ_{21} alone is expressed by the stress exposure $f_E^{(\tau_{21})}$, calculated with τ_{21} alone. The microdamage which is caused by both stresses, σ_2 and τ_{21}, acting simultaneously is expressed by the usual stress exposure f_E.

The influence of σ_2 on the $\tau_{21}(\gamma_{21})$-diagram and the influence of τ_{21} on the $\sigma_2(\varepsilon_2)$-diagram, respectively, can be expressed by the two semi-empirical equations (Eq. 127) and (Eq. 128) for the secant moduli $E_{\perp s}$ and $G_{\perp\parallel s}$:

$$E_{\perp s} = E_{\perp} - F_{\perp}(\sigma_2, \tau_{21})\left(E_{\perp} - E_{\perp s}\Big|_{f_E^{(\sigma_2)}=1}\right), \qquad \text{(Eq. 127)}$$

where $$F_{\perp}(\sigma_2, \tau_{21}) = \left(\frac{\left[f_E^{(\sigma_2)} + C^{(\tau_{21})}\left(f_E - f_E^{(\sigma_2)}\right)\right] - f_{E\,thr}^{(\sigma_2)}}{1 - f_{E\,thr}^{(\sigma_2)}}\right)^{n^{(\sigma_2)}}$$

$$G_{\perp\parallel s} = G_{\perp\parallel} - F_{\perp\parallel}(\tau_{21}, \sigma_2)\left(G_{\perp\parallel} - G_{\perp\parallel s}\Big|_{f_E^{(\tau_{21})}=1}\right), \qquad \text{(Eq. 128)}$$

where $$F_{\perp\parallel}(\tau_{21}, \sigma_2) = \left(\frac{\left[f_E^{(\tau_{21})} + C^{(\sigma_2)}\left(f_E - f_E^{(\tau_{21})}\right)\right] - f_{E\,thr}^{(\tau_{21})}}{1 - f_{E\,thr}^{(\tau_{21})}}\right)^{n^{(\tau_{21})}}$$

$E_{\perp s}$, $G_{\perp\|s}$ = actual secant moduli for a (σ_2, τ_{21}) state of stress

$E_{\perp s}\Big|_{f_E^{(\sigma_2)}=1}$; $G_{\perp\|s}\Big|_{f_E^{(\tau_{21})}=1}$ = secant moduli at IFF in an uniaxial σ_2-compression test or a pure τ_{21}-shear test, respectively

While f_E is calculated with σ_2, τ_{21} acting simultaneously, $f_E^{(\sigma_2)}$ and $f_E^{(\tau_{21})}$ are calculated with just one of the two combined stresses. From the limited amount of test data the following preliminary values can be recommended:

Table 10. Recommended values for parameters in (Eq. 127) and (Eq. 128)

		$f_{E\ thr}$	n	C
CFRP	τ_{21}	0.3	1.7	0.6
	σ_2	0.3	2.5	0.6
GFRP	τ_{21}	0.3	2.0	0.6
	σ_2	0.3	3.0	0.6

The comparison with experimental results shows that the model with its physical background is – with few parameters – very well capable of matching the non-linear stress/strain curves [Puck and Mannigel 2007].

The diagrams Fig. 98 and Fig. 99 show an example of the application of (Eq. 127) and (Eq. 128). Stress/strain-curves have been calculated for different ratios $\sigma_2/|\tau_{21}|$ which have also been used in experiments mentioned in [Puck and Mannigel 2007].

Nonlinearity due to changed fiber directions

In addition to the material non-linearities geometrical non-linearities (for instance in buckling problems) may occur in FRP-structures just as in structures made of isotropic material. A special form of non-linearity which is only relevant in FRP-structures is the change in fiber direction due to deformation of the structure under load.

In a thin walled laminated structure there is at least in good approximation a (σ_1, σ_2, τ_{21}) plane state of stress. A small quadratic element of the structure will be deformed to a rectangular shape by unequal strains ε_x and ε_y. Shear strain γ_{xy} will deform an originally quadratic element of the wall to a rhomb. This "angular deformation" may be partly achieved by a positive ('counter clockwise') rotation of the edge of the element which was originally parallel to the x-axis. The other part of the shear strain γ_{xy} is then

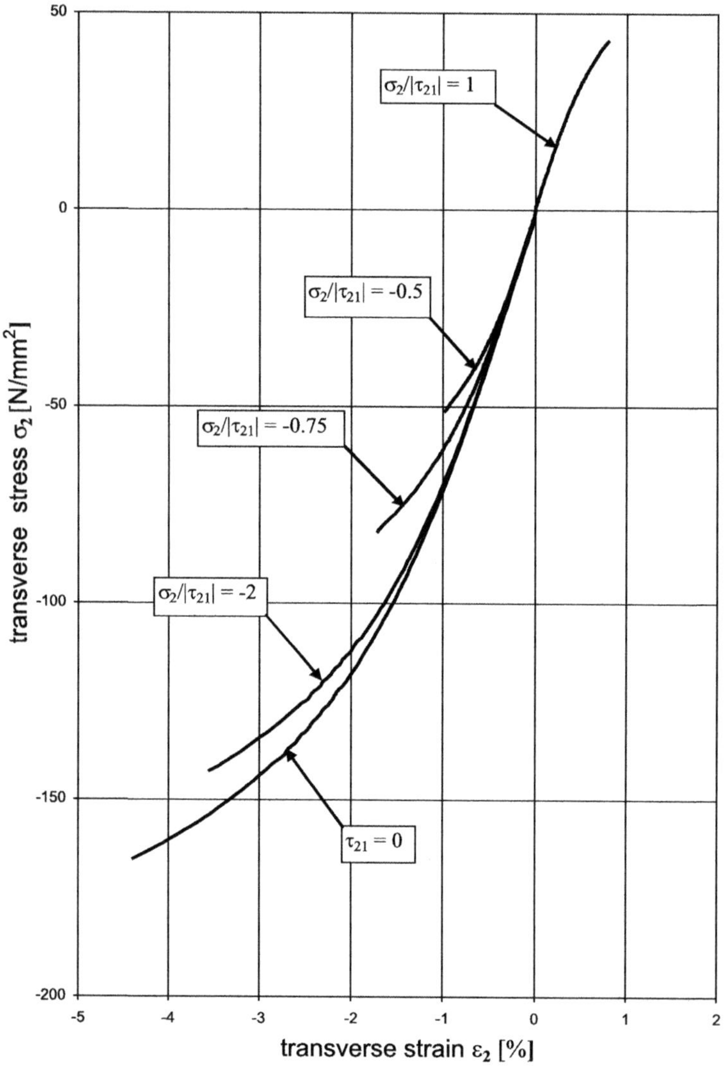

Fig. 98. $\sigma_2(\varepsilon_2)$ stress/strain-curves calculated from (Eq. 127)

added by a negative rotation of the element edge which was originally parallel to the y-axis. Under the simultaneously acting strains ε_x, ε_y, γ_{xy} a fiber with the original direction α will change its direction to $\alpha+\delta$ where δ can approximately be calculated by

$$\delta = \left(\varepsilon_y - \varepsilon_x\right)\sin\alpha\cos\alpha + \gamma_{xy}(p\cdot\cos^2\alpha - q\sin^2\alpha)$$
with $0 \le (p,q) \le 1 \wedge p+q=1$ (Eq. 129)

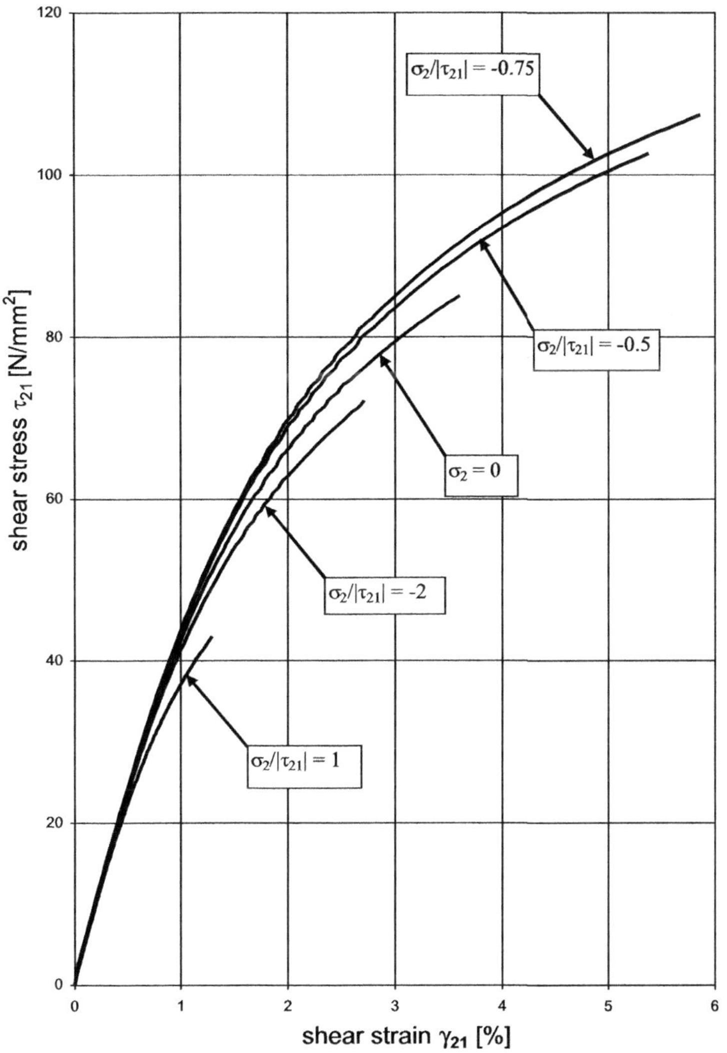

Fig. 99 $\tau_{21}(\gamma_{21})$ shear stress/strain-curves calculated from (Eq. 128)

If large shear strains γ_{xy} occur it is important to assume reasonable figures for p and q in (Eq. 129). These parameters depend on the boundary conditions of the element and these again depend on the boundary conditions and deformation behavior of the whole structure. For example, in a tubular drive shaft (with constant laminate lay-up over the whole length), which is fixed on the one end and loaded by a torsional moment at the other end, all cross sections will remain parallel to each other and to the fixed cross section at the one end. In this case – with x indicating the longitudinal direction and y indicating the circumferential direction – it would be correct to calculate δ with p = 1 and q = 0. A similar situation exists in a slender conical structure (for instance an aircraft wing or a rotor blade of a wind turbine). Quite different is the situation for a quadratic plate fixed in a shear frame on which a tension force is applied in the direction of the diagonal. Because of the symmetry of the deformation, p = q = 0.5 is correct here.

Summary of formulae for the Puck criteria in the regime of $\sigma_n < 0$

In the following the most important formula for the domain of $\sigma_n < 0$ (fracture modes B and C) are listed. In order to take into account the "strength-increasing" influence of σ_n Table 11 provides formulae both for the quadratic approach which Puck prefers and for the linear approach according to Coulomb's ideas. The latter has recently been preferred by some authors like Pinho [Pinho et al 2006]. The formulae for $R_{\perp\perp}^{A}$ are also used in the domain $\sigma_n \geq 0$ if not only stresses τ_{n1} but also stresses τ_{nt} do occur. Some additional remarks are made first on the preconditions for obtaining analytical relations for the stress exposure.

Preconditions for obtaining simple analytical relations for stress exposure f_E

The point of departure for calculating a stress exposure f_E will always be a *failure condition.* In the case of the UD lamina the fracture condition for IFF initially has the form of the 'master fracture condition' which is formulated in (σ_n, τ_{nt}, τ_{n1}) space:

$$F(\sigma_n, \tau_{nt}, \tau_{n1}, R_{\perp}^{t}, R_{\perp\perp}^{A}, R_{\perp\parallel}) = 1 \qquad \text{(Eq. 130)}$$

Frequently the F functions used for formulating failure conditions $F = 1$ are polynomials in which stress terms of the 1st and 2nd order occur. In such a case the stress exposure can be calculated as follows:

$$f_E = \frac{1}{2}\left(\sqrt{(\sum L)^2 + 4\sum Q} + \sum L\right) \qquad \text{(Eq. 131)}$$

where $\sum L$ = sum of the 1st order stress terms

and $\sum Q$ = sum of the 2nd order stress terms

On account of the modifications to the concepts of Coulomb and Mohr which are required to allow applicability to a UD lamina, the limits for IFF are first formulated in (σ_n, τ_{nt}, τ_{n1}) space and only after this transferred into (σ_1, σ_2, σ_3, τ_{23}, τ_{31}, τ_{21}) space.

The master fracture body in (σ_n, τ_{nt}, τ_{n1}) space in the half-space $\sigma_n > 0$ is closed because $\sigma_n > 0$ contributes to creating fracture and can even be the sole cause of the fracture; on the other hand, in the half-space $\sigma_n < 0$ it is open because compressive stresses σ_n can only serve to make fracture more difficult.

So, in the compression range $\sigma_n = \sigma_2 \cos^2 \theta_{fp}$ adds an additional fracture resistance to the fracture resistances $R^A_{\perp\perp}$ and $R_{\perp\parallel}$ of the action plane. This "strength-increasing" influence of σ_n can be formulated as a linear or a parabolic function. The fracture condition in the half-space $\sigma_n < 0$ will then be for the "parabolic" model:

$$\frac{\tau^2_{nt}}{(R^A_{\perp\perp})^2 - 2p^c_{\perp\perp}R^A_{\perp\perp}\sigma_n} + \frac{\tau^2_{n1}}{R^2_{\perp\parallel} - 2p^c_{\perp\parallel}R_{\perp\parallel}\sigma_n} = 1 \text{ for } \sigma_n < 0 \qquad \text{(Eq. 132)}$$

and for the "linear" model:

$$\left(\frac{\tau_{nt}}{R^A_{\perp\perp} - p^c_{\perp\perp}\sigma_n}\right)^2 + \left(\frac{\tau_{n1}}{R_{\perp\parallel} - p^c_{\perp\parallel}\sigma_n}\right)^2 = 1 \text{ for } \sigma_n < 0 \qquad \text{(Eq. 133)}$$

The linear model follows more closely the approach of Coulomb, while the parabolic model comes a little bit closer to the ideas of Mohr. With a mathematical manipulation – a "parameter coupling" – which is not physically justifiable but which nevertheless appears acceptable, it is possible to bring (Eq. 132) and (Eq. 133) into a form from which the stress exposure f_E can be calculated with the aid of equation (Eq. 131).

This is done with the following parameter coupling:

$$\frac{p^c_{\perp\perp}}{R^A_{\perp\perp}} = \frac{p^c_{\perp\parallel}}{R_{\perp\parallel}} = q \qquad \text{(Eq. 134)}$$

From this, with the quotient written q, we obtain the following from (Eq. 132) and (Eq. 133):

$$\left(\frac{\tau_{nt}}{R^{A}_{\perp\perp}}\right)^2 + \left(\frac{\tau_{n1}}{R_{\perp\|}}\right)^2 + 2q\sigma_n = 1 \text{ for } \sigma_n < 0 \qquad \text{(Eq. 135)}$$

$$\left(\frac{\tau_{nt}}{R^{A}_{\perp\perp}}\right)^2 + \left(\frac{\tau_{n1}}{R_{\perp\|}}\right)^2 - q^2\sigma_n^2 + 2q\sigma_n = 1 \text{ for } \sigma_n < 0 \qquad \text{(Eq. 136)}$$

With the help of the given equations it is possible to obtain the analytic relations shown in the table below for the parabolic as well as for the linear model. The decisive step here is the analytic calculation of the angle θ_{fp} of the fracture plane as a solution of the extremum problem $f_E(\theta) = f_E(\theta)_{max}$ from the condition $df_E/d\theta = 0$.

In [Pinho et al 2006] a linear model is presented for the consideration of the "strength-increasing" influence of σ_n. However the authors of [Pinho et al 2006] did not try to find an analytical solution as it is presented in the table below.

The parabolic model is preferred in Germany because in the range of high σ_n compressive stresses it delivers more conservative values for the supportable stresses τ_{nt}, τ_{n1} than does the linear model. With (σ_1, σ_2, τ_{21}) stress states the small difference between the results of the two models can be ignored for most applications.

It is not quite easy to conceive the exceptional universality of fracture conditions in the (σ_n, τ_{nt}, τ_{n1}) space. For the treatment of any arbitrary 3D-stress condition only two fracture criteria are necessary, one for $\sigma_n \geq 0$ and one for $\sigma_n < 0$, see Fig. 42. This is due to the unusual situation that in the fracture conditions formulated with σ_n, τ_{nt}, τ_{n1} the stresses are not fixed values, instead they vary with the fracture angle θ_{fp} which results from the solution of the extremum problem $f_E(\theta_{fp}) = f_{E\,max}$. If (Eq. 135) is applied to a (σ_2, τ_{21}) state of stress and $\theta_{fp} = 0\,°$, (Eq. 135) describes a piece of a parabola in the (σ_2, τ_{21}) plane (Mode B), see equation (P7) in Table 11, while (Eq. 136) describes a piece of a straight line. But if the (σ_2, τ_{21}) combination leads to an oblique fracture, $\theta_{fp} \neq 0$, surprisingly both (Eq. 135) and (Eq. 136) describe a piece of an ellipse (Mode C), see (L9) and (P9) in Table 11.

Table 11. The most important formulae for the domain of $\sigma_n < 0$ (fracture modes B and C) The formulae for $R^A_{\perp\perp}$ are also used in the domain $\sigma_n \geq 0$ if not only stresses τ_{n1} but also stresses τ_{nt} do occur.

	Linear model	No.	Parabolic model	No.
θ_{fp}	$\cos^2\theta_{fp} = \dfrac{R^A_{\perp\perp}}{R^c_\perp\sqrt{1+\left(p^c_{\perp\perp}\right)^2}}\left[\left(\dfrac{\tau_{21}}{R_{\perp\parallel}}\right)^2\left(\dfrac{R^A_{\perp\perp}}{\sigma_2}\right)^2+1\right]$ (*)	(L1)	$\cos^2\theta_{fp} = \dfrac{R^A_{\perp\perp}}{R^c_\perp}\left[\left(\dfrac{\tau_{21}}{R_{\perp\parallel}}\right)^2\left(\dfrac{R^A_{\perp\perp}}{\sigma_2}\right)^2+1\right]$	(*) (P1)
	$\cos^2\theta_{fp} = \dfrac{\left(-R^A_{\perp\perp}\right)}{\left(\sigma_2\right)_{fr}\sqrt{1+\left(p^c_{\perp\perp}\right)^2}}$ (*)	(L2)	$\cos^2\theta_{fp} = \dfrac{\left(-R^A_{\perp\perp}\right)}{\left(\sigma_2\right)_{fr}}$	(*) (P2)
$R^A_{\perp\perp}$	$R^A_{\perp\perp} = \dfrac{R^c_\perp}{2\sqrt{1+qR^c_\perp}}$ (*)	(L3)	$R^A_{\perp\perp} = \dfrac{1}{2q}\left(\sqrt{1+2qR^c_\perp}-1\right)$	(*) (P3)
	$R^A_{\perp\perp} = \dfrac{R^c_\perp}{2}\left(\sqrt{1+\left(p^c_{\perp\perp}\right)^2}-p^c_{\perp\perp}\right)$	(L4)	$R^A_{\perp\perp} = \dfrac{R^c_\perp}{2\left(1+p^c_{\perp\perp}\right)}$	(P4)
	$R^A_{\perp\perp} = \dfrac{R^c_\perp}{2\tan\Theta^c_{fp}}$	(L5)	$R^A_{\perp\perp} = R^c_\perp\cos^2\Theta^c_{fp}$	(P5)
$p^c_{\perp\perp}$	$p^c_{\perp\perp} = q\dfrac{R^c_\perp}{2\sqrt{1+qR^c_\perp}}$ (*)	(L6)	$p^c_{\perp\perp} = \dfrac{1}{2}\left(\sqrt{1+2qR^c_\perp}-1\right)$	(*) (P6)
(σ_2, τ_{21}) fracture condition for Mode B	$\tau_{21} = R_{\perp\parallel} - p^c_{\perp\parallel}\sigma_2$ (*)	(L7)	$\tau_{21} = \sqrt{R^2_{\perp\parallel} - 2\cdot p^c_{\perp\parallel}R_{\perp\parallel}\sigma_2}$	(*) (P7)

Table 11. Continued

	Linear model	No.	Parabolic model	No.
f_E for Mode B	$f_{\mathrm{E}} = \frac{1}{R_{\perp\parallel}}\left(\tau_{21} + p^{\mathrm{c}}_{\perp\parallel}\sigma_2\right)$ (*)	(L8)	$f_{\mathrm{E}} = \frac{1}{R_{\perp\parallel}}\left(\sqrt{\tau_{21}^2 + \left(p^{\mathrm{c}}_{\perp\parallel}\sigma_2\right)^2} + p^{\mathrm{c}}_{\perp\parallel}\sigma_2\right)$	(*) (P8)
(σ_2, τ_{21}) fracture condition for Mode C	$\frac{\tau_{21}^2}{\left(\sqrt{1+qR^{\mathrm{c}}_{\perp}}R_{\perp\parallel}\right)^2} + 4\left(\frac{\sigma_2}{R^{\mathrm{c}}_{\perp}}\right)^2 + 4\frac{\sigma_2}{R^{\mathrm{c}}_{\perp}} = 0$ (*)	(L9)	$\frac{\tau_{21}^2}{\left[\left(1+p^{\mathrm{c}}_{\perp\perp}\right)R_{\perp\parallel}\right]^2} + 4\left(\frac{\sigma_2}{R^{\mathrm{c}}_{\perp}}\right)^2 + 4\frac{\sigma_2}{R^{\mathrm{c}}_{\perp}} = 0$	(*) (P9)
$\tau_{21\max}$	$\tau_{21\max} = \sqrt{1+qR^{\mathrm{c}}_{\perp}}R_{\perp\parallel}$ at $\sigma_2 = -\frac{R^{\mathrm{c}}_{\perp}}{2}$ (*)	(L10)	$\tau_{21\max} = \left(1+p^{\mathrm{c}}_{\perp\perp}\right)R_{\perp\parallel}$ at $\sigma_2 = -\frac{R^{\mathrm{c}}_{\perp}}{2}$	(*) (P10)
$\left(\frac{\tau_{21}}{\sigma_2}\right)_{\mathrm{B/C}} = \arctan\zeta$	$\mp\left(\frac{R_{\perp\parallel}}{R^{\mathrm{A}}_{\perp\perp}}\sqrt{1+\left(p^{\mathrm{c}}_{\perp\perp}\right)^2} + p^{\mathrm{c}}_{\perp\parallel}\right)$ (*)	(L11)	$\mp\left(\frac{R_{\perp\parallel}}{R^{A}_{\perp\perp}}\sqrt{1+2p^{c}_{\perp\perp}}\right)$	(*) (P11)
f_E for Mode C	$f_{\mathrm{E}} = \frac{\tau_{21}^2\left(R^{\mathrm{A}}_{\perp\perp}\right)^2}{R^2_{\perp\parallel}(-R^{\mathrm{c}}_{\perp})\sigma_2} + \frac{\sigma_2}{(-R^{\mathrm{c}}_{\perp})}$ (*)	(L12)	$f_{\mathrm{E}} = \frac{\tau_{21}^2\left(R^{\mathrm{A}}_{\perp\perp}\right)^2}{R^2_{\perp\parallel}(-R^{\mathrm{c}}_{\perp})\sigma_2} + \frac{\sigma_2}{(-R^{\mathrm{c}}_{\perp})}$	(*) (P12)
'Weakening factor' η_{w1} for the effect of σ_1 on the IFF strengths $R^{\mathrm{t}}_{\perp}$, $R^{\mathrm{c}}_{\perp}$, $R_{\perp\parallel}$	$\eta_{\mathrm{w}_1} = \frac{c\left(a\sqrt{c^2(a^2-s^2)+1}+s\right)}{(ca)^2+1}$ where $c = \frac{f_{E_0}}{f_{E(FF)}}$ and $a = \frac{1-s}{\sqrt{1-m^2}}$ valid for $\frac{1}{s} \geq c \geq m$			(13)

Note: All equations marked with an asterisk (*) are not valid unless the parameter coupling $p^{c}_{\perp\perp} / R^{A}_{\perp\perp} = p^{c}_{\perp\parallel} / R_{\perp\parallel} = q$ is applied!

Notes on Formulae 1 to 13:

In general it is essential to realize that the strengths $R^{t}_{\perp}$, $R^{c}_{\perp}$, $R_{\perp\parallel}$ are "fixed" quantities which are material-specific and therefore do not depend on whether the linear or the quadratic model is used. The variables $R^{A}_{\perp\perp}$, $p^{c}_{\perp\perp}$, $p^{c}_{\perp\parallel}$ are, on the other hand, model-dependent variables - in other words, for the best possible analytic description of a measured (σ_2, τ_{21}) fracture curve different values $R^{A}_{\perp\perp}$ must, for example, be used depending on whether the linear or the parabolic model is used. The same applies to $p^{c}_{\perp\parallel}$ and $p^{c}_{\perp\perp}$.

Re (1) This formula can be applied if the stresses $(\sigma_2)_{fr}$ and $(\tau_{21})_{fr}$ at fracture are not yet known.

Re (2) This formula can only be used when $(\sigma_2)_{fr}$ at fracture is already known. If weakening due to σ_1 is already taking place, in the numerator $(-R^{A}_{\perp\perp})$ should be multiplied by the weakening factor η_{w1} according to equation (13). The stress σ_1 does not influence θ_{fp}.

Re (3) Normally the quotient $p^{c}_{\perp\parallel}/R_{\perp\parallel}$ is used here for q so as to take into account the parameter coupling $p^{c}_{\perp\perp} / R^{A}_{\perp\perp} = p^{c}_{\perp\parallel} / R_{\perp\parallel} = q$.

Re (4) These formulae apply globally - in other words, irrespective of whether parameter coupling is used or not.

Re (5) Θ^{c}_{fp} is the angle of the fracture plane occurring with uniaxial transverse compressive stress. The equations also apply without parameter coupling.

Re (6) Normally $q = p^{c}_{\perp\parallel} / R_{\perp\parallel}$ is used here to realize the parameter coupling $p^{c}_{\perp\perp} / R^{A}_{\perp\perp} = p^{c}_{\perp\parallel} / R_{\perp\parallel} = q$ with the $p^{c}_{\perp\perp}$ thus calculated.

Re (7) and (8) The only reason for taking the parameter coupling into consideration is to ensure that the fracture line for Mode B meets the ellipse for Mode C (in accordance with equation (9)) with equal slope ($-d\tau_{21}/d\sigma_2$) at the point of contact.

Re (9) The fracture condition is here described in a form which allows it to be seen that one is concerned with the equation of an ellipse which runs through the coordinates origin and through point ($-R^{c}_{\perp}$, 0). Even the magnitude of τ_{21max} can be obtained from the denominator of the τ_{21} term.

Re (10) Compare (9)

Re (11) In a (σ_2, τ_{21})-fracture diagram the angle 2ζ marks the sector, in which (σ_2, τ_{21})-vectors will lead to an IFF of mode C.

Re (12) Equations (L12) and (P12) for the stress exposure f_E in Mode C are formally completely identical for both models, however the values for $R^{A}_{\perp\perp}$ are different!

Re (13) In the range of validity of (13) the stress exposures calculated for Modes A, B and C using equations (L8), (P8), (L12), (P12) must be increased by dividing f_{E_0} by η_{w1}. In the formula for the weakening factor η_{w1}, f_{E_0} means the IFF stress exposure without σ_1 influence; $f_{E(FF)} = \sigma_1 / R_{\parallel}^{t}$ for $\sigma_1 > 0$ or $f_{E(FF)} = \sigma_1 / (-R_{\parallel}^{c})$ for $\sigma_1 < 0$ is the fibre-fracture stress exposure. $sR_{\parallel}^{t}$ or $s(-R_{\parallel}^{c})$ is the stress σ_1 at which weakening due to σ_1 begins to become noticeable. The remaining minimum value of the IFF strengths in the form of $mR_{\perp}^{t}$, $mR_{\perp}^{c}$ or $mR_{\perp\parallel}$ are given by the fraction m which applies when σ_1 is theoretically equal to $R_{\parallel}^{t}$ or $(-R_{\parallel}^{c})$. In the absence of test results it is recommended that calculations use $s = m = 0.5$. Different values of (s, m) can be used for $\sigma_1 > 0$ and $\sigma_1 < 0$.

Terms and definitions

Abbreviations

A, B, C	IFF-Fracture modes
CFRP	Carbon-fiber-reinforced plastic
CLT	Classical Laminate Theory
COS	Coordinate system
Eq.	Equation
FF	Fiber Fracture
FEM	Finite-Element-Method
FRP	Fiber-reinforced plastic
GFRP	Glass-fiber-reinforced plastic
IFF	Inter Fiber Fracture
MF	Margin of Safety
PA	Polyamide
PEEK	Polyetheretherketon
RF	Reserve Factor
UD	unidirectional
2D	two-dimensional
3D	three-dimensional

Symbols

$\parallel$, $\perp$	parallel to fiber and transverse (perpendicular) to fiber direction

Latin characters

$E_{\parallel}$, $E_{\perp}$	Young's Modulus of UD lamina for parallel or transverse to fiber direction
$E_{\perp s}$,	secant Young's modulus of UD lamina at IFF-initiation
c	parameter for the calibration of the degradation curve
f_E	stress exposure
$f_{E,\ FF}$	stress exposure Fiber Fracture
$f_{E,\ IFF}$	stress exposure Inter Fiber Fracture
f'_E	standardized stress exposure
$f_S = f_E^{-1}$	stretch factor
f_S^L	stretch factor of load determined stresses
$G_{\perp\parallel}$	in-plane shear modulus of UD lamina
$G_{\perp\parallel s}$	secant shear modulus of UD lamina at IFF initiation
m	magnification factor
N	Force
p, q	parameters for the calculation of the change in fiber direction under load
$p_{\perp\parallel}^t$, $p_{\perp\parallel}^c$	inclination of (σ_2, τ_{21})-fracture curve at $\sigma_2 = 0$; t for the range $\sigma_2 > 0$; c for the range $\sigma_2 < 0$
$p_{\perp\perp}^t$, $p_{\perp\perp}^c$	inclination of (τnt, σn)-fracture curve at $\sigma_n = 0$; t for the range $\sigma_2 > 0$; c for the range $\sigma_2 < 0$
$R_{\parallel}^t$, $R_{\parallel}^c$	tension and compression strength of UD lamina parallel to fiber direction
$R_{\perp\parallel}$	in-plane shear strength of lamina
$R_{\perp}^t$, $R_{\perp}^c$	tensile and compressive strength of UD lamina transverse to fiber direction
$R_{\perp\perp}^A$	fracture resistance of an action-plane parallel to the fiber direction against its fracture due to $\tau_{\perp\perp}$ stressing acting on it
S	sum

Greek characters

α	angle of fiber direction (right-hand COS, positive from x to x_1 anticlockwise)
α_{f_E}, α_E, α_G	degradation parameters
δ	crack density, deviation angle
ε	strain
η	reduction factor of the elasticity values after exceeding the IFF limit

η_E, η_G, η_ν	reduction factor for Young's modulus E_2, shear modulus G_2 and Poisson's ratio ν_{21} (major Poisson's ratio) respectively
η_{m+p}	correction factor for micro-damage and probabilistics
η_{w1}	degradation factor, weakening due to σ_1
φ	fiber-volume fraction, angle
θ	angle between thickness direction and a parallel-to-fiber section plane of the UD lamina
θ_{fp}	angle of the fracture plane
ν	Poisson's ratio (for isotropic case)
$\nu_{\perp\parallel}$, $\nu_{\parallel\perp}$, $\nu_{\perp\perp}$	Poisson's ratios of UD lamina
$\nu_{\perp\parallel s}$	major Poisson's ratio of UD lamina at IFF-initiation
ψ	angle calculated from arctan (τ_{n1}/τ_{nt}) determined by the ratio τ_{n1}/τ_{nt}
σ_1, σ_2, σ_3	normal lamina stresses related to the local lamina COS
σ_n	normal stress on an inclined action plane
$\sigma_\parallel$	longitudinal stressing in the UD-lamina
$\sigma_\perp$	transverse stressing in the UD-lamina
$\{\sigma\}_{m+p}$	fracture stress vector taking (m+p) effects into account,
$\{\sigma\}_{fr}^{Mohr}$	fracture stress vector without taking (m+p) effects into account
τ_{xy}	in-plane shear stresses of the laminate
τ_{zy}, τ_{zx}	out-of-plane shear stresses of the laminate
τ_{21}, τ_{23}, τ_{32}	shear lamina stresses related to the local lamina COS
τ_{nt}	transverse shear stress on an inclined action plane (action-plane strength criterion)
τ_{n1}	longitudinal shear stress on an inclined action plane
$\tau_{\perp\parallel}$	transverse/longitudinal shear stressing of the UD-lamina
$\tau_{\perp\perp}$	transverse/transverse shear stressing of the UD-lamina
γ	shear strain
ω	damage parameter; half of the crossing angle of an angle ply laminate, measured from the x-axis: $\alpha_1 = +\omega$, $\alpha_2 = -\omega$
ξ	parameter for the calibration of the degradation curve
Δ	difference
Δ_{max}	relative difference between the fracture stress vector calculated according to Mohr and that corrected by η_{m+p}

Laminate notation

$[0/90]_s$	definition of a symmetrical laminate with respect to the mid-plane

Superscripts

'	standardized, slope
A	action plane related
c	compressive
comb	combined
L	load-determined
r	residual
t	tensile

Subscripts

cor	correction
f	fiber
fr	fracture
k	k^{th} lamina
m+p	micro-damage and probabilistic
ref	reference
s	stretch
w	weakening

Coordinate systems

x, y, z	laminate COS
x_1, x_2, x_3	lamina COS
x_I, x_{II}, x_{III}	directions of principal normal stresses
$\parallel$, $\perp$	UD-lamina COS showing the parallel and transverse fiber directions (for denoting transversely-isotropic properties)
n, t	normal, tangential to a parallel to the fiber plane of the lamina

References

Bailey JE, Curtis PT, Parvizi A (1979) On the transverse cracking and longitudinal splitting behavior of glass and carbon fiber reinforced epoxy cross ply laminates and the effect of Poisson and thermally generated strain, Proc. R. Soc. Lond. A 366, 599–623

Boniface L, Smith PA, Rezaifard HA, Bader MG (1997) Transversely ply cracking in composite laminates – initiation or propagation controlled?, Journal of Composite Materials 31, 1080–1112

Busse G, Kröplin B-H, Wittel F (2006) Damage and its evolution in fiber composite materials: Simulation and non-destructive evaluation, Universität Stuttgart, ISBN: 3930683903

Chang FK, Chang KY (1987) A progressive damage model for laminated composites containing stress concentrations, Journal of Composite Materials 21, 834 ff

Coulomb CA (1776) Sur une Application des Règles de Maximis et Minimis a quelques Problèmes de Statique, relatifs a l´Architecture (French), Mémoires de Mathématiques et de Physique, Académie de Royal des Sciences par divers Savans, Paris

Cuntze R, Kopp JW, Huybrechts D (1997) Neue Bruchkriterien und Festigkeitsnachweis für unidirektionalen Faserkunststoffverbund unter mehrachsiger Beanspruchung – Modelle und Experimente, VDI-Fortschrittberichte, Reihe 5: Grund und Werkstoffe, VDI-Verlag, Düsseldorf, German

Daniel IM, Ishai O (1994) Engineering Mechanics of Composite Materials, Oxford University Press, New York/Oxford

Deuschle M, Kröplin B-H (2008) FE Implementation of Puck's failure theory for fibre reinforced composites under 3D-stress, submitted to Composites Science and Technology for publication in the World Wide Failure Exercise No2

Det Norske Veritas (2003), DNV-RP-F202 Composite Risers

Edge EC (1998) Stress based grant-sanders method for predicting failure of composite laminates, Composites Science and Technology 58, 1033–1044

Fischer O (2003) Fiber fracture behavior in fiber reinforced plastics, Rheinisch-Westfälische Technische Hochschule Aachen, Ph.D.-thesis, German

Flaggs DL, Kural MH (1982) Experimental Determination of the In Situ Transverse Lamina Strength in Graphite Epoxy Laminates, Journal of Composite Materials, Vol. 16, 103–116

Garret KW, Bailey JE (1977) Multiple transverse fracture in 90° cross ply laminates of a glass fiber-reinforced polyester, Journal of Material Science 12, 157–168

Germanische Lloyd, (2003) Guideline for the Certification of Wind Turbines, Edition 2003 with Supplement 2004

Gotsis PK, Chamis CC, Minnetyan L (1998) Prediction of composite laminate fracture: Micromechanics and progressive fracture, Composites Science and Technology, Vol. 58, 1137–1149

Groves SE, Harris CE, Highsmith AL, Allen DH, Novell RG (1987) An experimental and analytical treatment of matrix cracking in cross-ply laminates, Experimental Mechanics, 73–79

Hart-Smith LJ (1984) Approximate analysis methods for fibrous composite laminates under combined biaxial and shear loading, Douglas Aircraft Company, IRAD Report MDC-J9898

Hart-Smith LJ (1998) Predictions of a generalized maximum-shear-stress failure criterion for certain fibrous composite laminates, Composites Science and Technology, special issue, 58: 1179–1208

Hart-Smith LJ (1998) Predictions of the original and truncated maximum strain failure models for certain fibrous composite laminates, Composites Science and Technology, special issue, 58: 1151–1178

Hashin Z (1980) Failure criteria for unidirectional fiber composites, Journal of Applied Mechanics 47: 329–334

Hashin Z (1985) Analysis of cracked laminates: a variational approach, Mechanics of Materials 4, 121–136

Highsmith AL, Reifsnider KL (1982) Stiffness-reduction mechanisms in composite laminates. Damage in composite materials ASTM-STP-775, American Society for Testing of materials, 103–117

Hinton MJ, Kaddour AS., Soden PD (2002) A comparison of the predictive capabilities of current failure theories for composite laminates, judged against experimental evidence. Composites Science and Technology 62: 1725–1797

Hodgkinson E (1861) Experimental researches on the strength on other properties of cast iron Practical Essays on the Strength of Cast Iron, 5th edition by T. Tredgold, edited by Hodgkinson, London, England

Huybrechts D (1996) Ein erster Beitrag zur Verifikation des wirkebenebezogenen Zwischenfaserbruchkriteriums nach Puck (a first contribution in verifying the action-plane-related inter-fiber-failure criterion after Puck), Technisch-wissenschaftlicher Bericht, IKV Bd. 44, Verlag Mainz, Aachen

Johnson JB (1926) Johnsons´s materials of construction 6th edition, edited by F.E. Turneaure, John Wiley & Sons, Inc. New York, USA

Kadatoni F, Aki F (1984) Analysis of the interlaminar shear strength of Mica/Epoxy Insulations, Composites 15 1, 57 ff

Kaiser Kuhnel Obst (2004) Failure criteria for non-metallic materials, Part I: fibre reinforced plastics. In: Proceedings of the 11th European Conference on Composite Materials, Rhodos, Greece

Knickrehm A (2000) Zum Versagen unidirektionaler Glasfaser-Kunststoff-Verbunde bei Biegeschwellbeanspruchung (German), Verlag Shaker, ISBN 3-8265-7025-1

Knight NF, Rankin CC, Brogan FA (2002) STAGS computational procedure for progressive failure analysis of laminated composite structures, International Journal of Non-Linear Mechanics, 833–849

Knops M (2003) "Sukzessives Bruchgeschehen in Faserverbundlaminaten" (Gradual failure process in fiber/polymer laminates), technisch wissenschaftlicher Bericht, IKV Aachen, Band 140, Verlag Mainz, Aachen, ISBN 3-86130-480-5, German

Knops M, Bögle C (2006) Gradual failure process in fiber/polymer laminates, Composites Science and Technology, special issue on Composite Materials Reliability and life prediction of composite structures, Volume 66, Issue 5, 616–625

Kopp JW (1998) More efficient composite component design using action-plane related IFF strength criteria, Proceedings 43rd International SAMPE Conference, Anaheim

Kopp JW (2000) Zur Spannungs- und Festigkeitsanalyse für unidirektionalen Faserverbundkunststoff (Contribution to the Stress and Strength Analysis of Unidirectionally Fiber Reinforced Plastics), Technisch-wissenschaftlicher Bericht, IKV Bd. 99, Verlag Mainz, Aachen

Kopp JW, Michaeli W (1999) The new action plane related strength criterion in comparison with common strength criteria, In: Proceedings ICCM12, Paris

Kopp JW, Puck A and Michaeli W (1997) Modeling and Experiments for a Physically Based Strength Hypothesis for Fiber-Matrix-Laminates (German), Proceedings of the DGM-Conference, Kaiserslautern, Germany, 751–758.

Kuhnel E (2008) Bewertung der Anwendbarkeit von Zwischenfaserbruch-Kriterien auf endlosfaserverstärkte Kunststoffe mit thermoplastischer Matrix (Evaluation of the applicability of interfibre failure criteria on endless fibre reinforced plastics with thermoplastic matrix), Technisch-wissenschaftlicher Bericht, Verlag Mainz, Aachen

Lambrecht L (2007) Numerische Experimente zur Analyze der Einflussgrößen auf die Degradation von mehrschichtigem GFK-Laminat (German), Diplomarbeit am IKV Aachen, Betreuer: E. Kuhnel

Lavoie A, Morton J (2000) LAVOIE Scaling of first-ply failure and strength in $[+\theta_n/-\theta_n/90_{2n}]_s$ laminates: Experiments and predictions, Journal of Composites Technology & Research, Vol. 22, No. 3, 153–160

Li S, Reid SR, Soden PD (1994) A finite strip analysis of cracked laminates, Mechanics of materials 18, 289–311

Li S, Reid SR, Soden PD (1998) A continuum damage model for transverse matrix cracking in laminated fiber-reinforced composites, Phil. Trans. R. Soc. London. A 356, 2379–2412

Li S, Reid SR, Soden PD, Hinton MJ (2000) The nonlinear structural response of laminated composites, Crown DERA, 1–10

Liu KS, Tsai SW (1998) A Progressive quadratic failure criterion for a laminate – Part A, Composites Science and Technology, Vol. 58, 1023–1032

Lutz G (2006) The Puck theory of failure in laminates in the context of the new guideline VDI2014 Part 3, Proceedings Conference on Damage in Composite Materials, Stuttgart, (available online at http://www.ndt.net/v11n12.htm)

Mannigel M (2007) "Einfluss von Schubspannungen auf das Faserbruchgeschehen in kohlenstofffaserverstärkten Kunststoffen (CFK)" (Influence of shear stresses on the fiber failure behavior in carbon fiber reinforced plastics (CFRP)) technisch wissenschaftlicher Bericht, IKV Aachen, Band 183, Verlag Mainz, Aachen, ISBN 3-86130-853-3, German

Mc Cartney LN (1998) Prediction of transverse crack formation in cross-ply laminates, Composites Science and Technology 58, 1069–1081

Mc Cartney LN (2002) Prediction of ply crack formation and failure in laminates, Composites Science and Technology 62, 1619–1631

Michaeli W, Knops M, Fischer O, Eifel D (2002) Investigations on the fiber fracture behavior of carbon fiber reinforced plastics, In: Proceedings Sampe, Long Beach

Mohr O (1900) Welche Umstände bedingen die Elastizitätsgrenze und den Bruch eines Materials? (German), Zeitschrift des Vereins deutscher Ingenieure 24: 1524 ff.

Nairn JA, Hu S, Bark JS (1993) A critical evaluation of theories for predicting microcracking in composite laminates, Journal of Materials Science 28, 5099–5111

Nairn JA (2000) Matrix Microcracking in Composites, Polymer Matrix Composites, Chapter 13, Volume 2 of Comprehensive composite materials

Nairn JA, Hu S (1992) The formation and effect of outer-ply microcracks in cross-ply laminates: a variational approach, Engineering Fracture Mechanics 41, 203–221

Nairn JA, Hu S (1994) Matrix microcracking Damage Mechanics of Composite Materials, Elsevier Science B.V., 187 ff

Liu KS, Tsai SW (1998) A Progressive quadratic failure criterion for a laminate – Part A. Composites Science and Technology 58: 1023–1032

Paul B (1961) A Modification of the Coulomb-Mohr Theory of Fracture, Journal of Applied Mechanics, 259–268

Pinho ST, Iannucci L, Robinson P (2006) Physically-based failure models and criteria for laminated fibre-reinforced composites with emphasis on fibre kinking: Part I: Development, Composites A, 37 (1): 63–73

Puck A (1967) Zur Beanspruchung und Verformung von GFK-Mehrschichtverbundbauelementen (German), Kunststoffe (German Plastics) 57 12, 965 ff

Puck A (1969) Calculating the strength of glass fiber/plastic laminates under combined load. Kunststoffe, German Plastics 55: 18 ff.

Puck A. (1969) Festigkeitsberechnung an Glasfaser/Kunststoff-Laminaten bei zusammengesetzter Beanspruchung; Bruchhypothesen und schichtweise Bruchanalyse ("Strength analysis on GRP-laminates under combined stresses; fracture hypotheses and layer-by-layer failure analysis"), Kunststoffe, German Plastics 59 (bilingual edition English and German), 18–19, German text 780–787

Puck A. (1992) Designing FRP-laminates with strain or stress criteria? Faser-Kunststoff-Verbunde mit Dehnungs- oder Spannungskriterien auslegen?, Kunststoffe, German Plastics 82 (bilingual edition English and German), 149–155

Puck A. (1992) Ein Bruchkriterium gibt die Richtung an ("A fracture criteria shows the direction"), Kunststoffe, German Plastics 82 (5) (bilingual edition English and German), German text 207–210

Puck A (1996) Festigkeitsanalyse von Faser-Matrix-Laminaten – Modelle für die Praxis (German). Carl Hanser, München, Wien; download from http://www.klub.tu-darmstadt.de/forschung/download.php

Puck A. (1997) "Physically based IFF-criteria allow realistic strength analysis of fiber-matrix-laminates" (German), Proceedings of the DGLR-Conference 1996, Ottobrunn, Germany, 1996, 315–352

Puck A, Deuschle M (2008) A simple physically based degradation model for FRP-UD laminae after onset of cracking caused by inter fibre fracture, submitted for publication as a short note to Composites Science and Technology

Puck A., Mannigel, M. (2007) Physically based stress-strain relations for the Inter Fiber Fracture analysis of FRP laminates, Composites Science and Technology, Volume 67, Issue 9, July 2007, 1955–1964

Puck A, Schneider W (1969) On failure mechanisms and failure criteria of filament-wound glass-fiber/resin composites. Plastics & Polymers, The plastics institute transactions and journal, Pergamont Press, Oxford: 33–42

Puck A., Schürmann H. (1998) Failure analysis of FRP Laminates by means of physically based phenomenological models, Comp. Sci. and Techn. 58, 1045–1067

Puck A., Schürmann H. (2002) Failure analysis of FRP Laminates by means of physically based phenomenological models, Comp. Sci. and Techn. 62, 1633–1662

Puck A, Garbe (1993) Erfahrungen mit Bruchkriterien an schwellend belasteten GFK-Drehrohrfedern, Kunststoffe 83, 406–411

Puck A, Kopp J, Knops M (2002) Guidelines for the determination of the parameters in Puck's action plane strength criterion, Composites Science and Technology 62, 371–378

Ritt S. (1999) Development of a concept and programming of algorithms for the visualisation of the 'blind spots' on the fracture body surface of Puck's action plane related Inter Fiber Failure criterion (German), Institute of Plastics Processing, Technical University of Aachen, Diploma Thesis – supervisors: J. Kopp, M. Knops

Schürmann, H. (1989) Zur Erhöhung der Belastbarkeit von Bauteilen aus Faser-Kunststoff-Verbunden durch gezielt eingebrachte Eigenspannungen, Dissertation an der Universität-Gesamthochschule Kassel, Fortschrittsbericht Reihe 1, Nr. 170, VDI-Verlag, Düsseldorf

Schürmann, H. (2004) Konstruieren mit Faser-Kunststoff-Verbunden (German). Springer-Verlag, Berlin, Heidelberg, New York, 398

Smith PA, Boniface L, Glass NFC (1998) A comparison of transverse cracking phenomena in (0/90)s and (90/0)s laminates, Applied Composite Materials 5, 11–23

Smith PA, Wood JR (1990) Poisson's ratio as a damage parameter in the static tensile loading of simple crossply laminates, Composites Science and Technology 38, 85–93

Soden PD, Hinton MJ, Kaddour AS (1998) A comparison of the predictive capabilities of current failure theories for composite laminates. Composites Science and Technology 58: 1225–1254

Soden PD, Hinton MJ, Kaddour AS (2002) Biaxial test results for strength and deformation of a range of E-glass and carbon fiber reinforced composite laminates: failure exercise bench-mark data, Composites Science and Technology 62, 1489–1514

Surgeon M, Vanswijgenhoven E, Wevers M, van der Biest O (1999) Transverse cracking and Poison's ratio reduction in cross-ply carbon fiber-reinforced polymers, Journal of Materials Science, 5513–5517

Talreja R (1994) Damage mechanics of composite materials, Elsevier, Amsterdam

Thionnet A, Rennard J (1993) Meso-macro approach to transverse cracking in laminated composites using Talreja's model, Composites Engineering 3, 851–871

Tsai CL, Daniel IM (1992) The behavior of cracked cross-ply composite laminates under shear loading, International Journal of Solids Structures 29, No. 24, 3251–3267

Tsai SW, Wu E (1971) A General Theory of Strength for Anisotropic Materials. Journal of Composite Materials 5: 58 ff

Tsai SW (1992) Theory of Composites Design, THINK Composites, ISBN 0-9618090-3-5, 9.5–9.6

VDI 2014 Blatt 3 (2006) VDI-Richtlinie 2014 Blatt 3: Entwicklung von Bauteilen aus Faser-Kunststoff-Verbund, Berechnung, Ausg. deutsch/englisch (VDI Guideline 2014 part 3: Development of Fibre-Reinforced-Plastics components, Analysis, Issue German/English), Beuth-Verlag, Berlin (2006)

Whitney JM (2000) Effective elastic constants of bidirectional laminates containing transverse ply cracks, Journal of Composite Materials, 2000, 955–978

Wittel K, Kun F, Kröplin BH, Herrmann HJ (2001) A study of transverse ply cracking using a discrete element method. Computational Materials Science: 608–619

Wolf E, Butalis TS (1998) A strain-energy based failure criterion for non-linear analysis of composite laminates subjected to biaxial loading, Composites Science and Technology 58, 1107–1124

Index

GPSR Compliance

The European Union's (EU) General Product Safety Regulation (GPSR) is a set of rules that requires consumer products to be safe and our obligations to ensure this.

If you have any concerns about our products, you can contact us on ProductSafety@springernature.com

In case Publisher is established outside the EU, the EU authorized representative is:

Springer Nature Customer Service Center GmbH
Europaplatz 3
69115 Heidelberg, Germany

Batch number: 10755292

Printed by Printforce, the Netherlands